Pol Chantraine

THE LIVING ICE

The Story of the Seals and the Men
Who Hunt Them in the Gulf of St. Lawrence

Translated by David Lobdell

McCLELLAND AND STEWART

Copyright © 1980 by McClelland and Stewart Limited

All Rights Reserved

The Canadian Publishers
McClelland and Stewart Limited
25 Hollinger Road
Toronto   M4B 3G2

CANADIAN CATALOGUING IN PUBLICATION DATA

Chantraine, Pol, 1944 -
   The living ice

ISBN 0-7710-1960-2

1. Sealing – Quebec (Province) – Magdalen Islands.
2. Sealing – History.   3. Seals (Animals).
4. Animals, Treatment of.   I. Title.

SH363.C53   639′.29′09714797   C79-094884-2

Photographs and Maps by Ronald Labelle

This translation was completed with the support of The Canada Council

Printed and bound in Canada
by John Deyell Company

Contents

I   The Living Ice *7*
II   Nazaire *19*
III   The First Hunt *32*
IV   The Art of Sealing *41*
V   Days of Fever *52*
VI   The Odyssey of the Magdaleners *73*
VII   The End of the Walrus *88*
VIII   The Pagophiles *103*
IX   Of Seals and Men *117*
X   The Great Hunts *133*
XI   Steam and the Decline *148*
XII   Other Times, Other Seals *160*
XIII   Springs of Mourning *173*
XIV   The Disasters *185*
XV   The Great Seal Herd *204*
XVI   Seals and Captivity *216*
       Epilogue *227*
       Notes *237*

for Marc à Donat

I
The Living Ice

"This here," said Wellie, "is the shore ice. Down there is the living ice."

With a vague gesture, he indicated a great expanse of ocean, invisible from our vantage point, that lay somewhere beyond the hummocks screening the horizon a mile or so before us.

The living ice. It was the first time I had heard that expression, but I could appreciate its aptness, although we are more inclined to associate the cold, desolate expanses of the ice-pack with death than life.

Wellie moved on, up to his knees in snow, constantly scanning his surroundings, inhaling deeply, sniffing the air. At one moment, his head was raised, his nostrils flared; at the next, his eyes were lowered, his ears pricked, an anxious look on his face – probing nature with all his senses, out of habit and instinct, for it is from her that he derives the greater part of his sustenance.

We had just raised a smelt net from a small, quiet pond surrounded by ice at the mouth of one of the mountain's streams, and we were getting our muscles toned for the hunt by making our way along the *débarris* or shore ice that lay all about us flat and immobile, riveted to the shore and stranded on the reefs, prisoner of the coves and capes. Farther out lay the living ice. Slabs, pans, and floes of various sizes and consistencies floated freely in the currents, whirling, colliding, jostling, raftering, crumbling, at the mercy of winds, waves, temperatures – a crazy, chaotic procession of ice ebbing and flowing in the rising and falling tides.

Roving, drifting, this ice is alive, not only in the sense that it moves, rather like blood corpuscles seen under a microscope, but also in that it bears life, a great flood of life that has been the livelihood of humans ever since they first settled on the shores of the Gulf: watching men come and go, the victims of other men and other civilizations or of their own vain thirst for conquest, a flood of life that, long after man has vanished from the scene, will continue to exist, adrift on the floating ice, indifferent to time and the events transpiring about it – the seals.

This icescape is in constant motion, as is the life that makes its home on it – from the pelagian plankton that thrives here due to the greenhouse effect of the sun's rays striking the ice, to the seals and the men who hunt them, all bound together in a complex network of trophic bonds, an ineluctable chain of predators and prey engaged in a mad scramble for survival, the supreme arbitrator of which is Time.

Northern, subarctic, neo-polar – there is no lack of adjectives to describe this universe of frozen sea-water that is our home four months of the year, with its celebrated climatic extremes; but far from doing justice to the almost meridional geographic location of our little archipelago – 47° north latitude – the most they can do is boggle the minds of eighty per cent of the human race with abstract, exotic notions of the kingdom of ice.

Here, the sea generally freezes over about mid-January.

It begins with a thin rime that covers the surface of the water. Next a sort of fleecy evaporation rises from water the colour of molten lead ("The seas are heavy," the fishermen say); and then, almost overnight, with the first sharp drop in temperature, this veil of fog condenses into small lumps of ice that merge to form larger blocks. Tossed about by the waves, these blocks adhere to each other, resulting in slabs several meters in diameter, which in turn fuse to form pans, floes, and finally the ice-field itself.

In its early stages, this ice is soft, fragile and green in colour, a green so pure as to seem almost unreal (salt-water ice remains somewhere between the liquid and solid states as long as the sea about it steams); but soon the wind gives it a sheen, compresses and packs it down, and it begins to solidify. It is also buffeted by rain, snow, foam, spray, all of which congeal the moment they touch it (sea-water freezes at about −8°C, depending on its saltiness), adding to its thickness and solidity.

At the mercy of the waves and winds, this indigenous new ice, which appears simultaneously throughout almost the entire area, comes into contact with the harder, fresh-water ice that the ma-

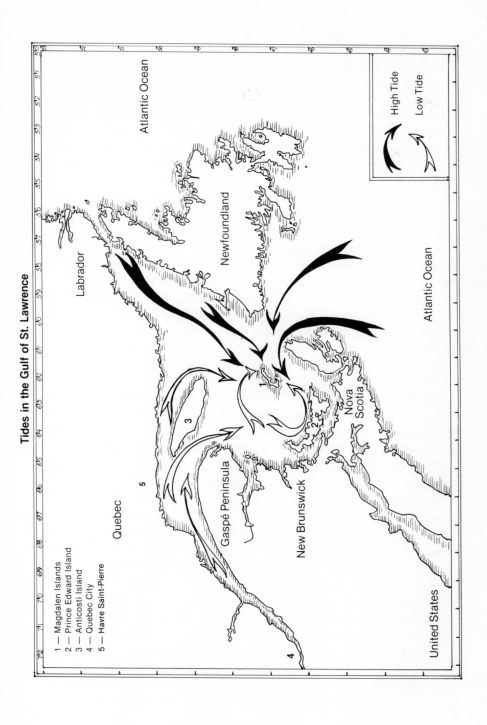

jestic St. Lawrence tirelessly disgorges and which is pulled diagonally across the Gulf from Gaspé to the Cabot Strait by the surface currents of the Laurentian Channel. Farther north, other indigenous ice meets up with the polar ice-floe, which penetrates the Gulf with each rising tide by way of the Strait of Belle Isle. This is old ice, sometimes several years old, and is as hard as granite and a deep metallic blue in colour. It is carried from the Arctic by the cold Labrador current, the major determining factor in our climate.

Tempered by wind and cold, these separate bodies of ice coalesce to generate more and more new ice, until almost the entire surface of the Gulf is covered by it. Only two narrow channels remain open to navigation: one along the North Shore, the other along the west coast of Newfoundland where the currents are strong enough to clear the ice away as quickly as it is formed. On a fine day, when you take off in a plane for Gaspé, all you can see beneath you, for nearly two hundred miles, is a dazzling white sheet of ice with a maze of giant cracks crisscrossing it, an ice-field of vast circular or oval floes that resemble lunar craters with their hummocky ranges, interspersed with long, jagged water leads in which the seawater shimmers and gleams.

Storms throw this unreal landscape into total chaos. Gusts of up to seventy knots, even stronger in mid-winter, hurl the floes against each other. With a terrible din, they collide, capsize, and straddle each other; and the spectacle they offer once the cataclysm has passed bears a striking resemblance, all things considered, to a country devastated by a major earthquake.

The edges of the floes, which receive the brunt of the shock, resemble raised buttresses; these are the hummocks, or *bouscueuils* as the Magdaleners call them. Almost everywhere you look, there are pressure ridges – miniature mountain chains composed of chunks of ice and rarely higher than a man's shoulders – where the edges of the ice-floes have raftered. Elsewhere, enormous blocks of ice stand stacked upon one another, while statuesque sculptures of translucent ice prisms moulded on the spot by the hands of Jack Frost repose in a variety of outlandish postures, some of which actually seem to defy gravity. Long sharp needles, arrows, and obelisks raise their spiky profiles to the sky, and highly intricate structures climb in tiers to their lofty summits, then descend gently back to the ice bed, all covered with a layer of snow packed to the density of steel by the howling wind. There are plains, valleys, deep polished craters lined with vast sheets of shimmering ice, crevasses, streams of frosy (a mixture of ice shavings that only appears to be solid), and patches of clear water, called *saignées* or leads, where large slabs and pans bob lazily in the current, the playground of the seals.

But the seals were too far away – perhaps as many as thirty or forty miles from the archipelago – to allow Wellie and I even to dream of seeing them that day. Nonetheless, our thoughts were projected in that direction, while we continued to make our way along the shore ice, joyfully savouring the silence and the cold sea air that stretched all about us. And, with all due respect to those whom the media have induced into thinking otherwise, there was a real sense of affection in our thoughts, knowing that this soft snow that hampered our progress would provide a choice breeding-ground for the new generation of pagophiles[1] that the female seals were already giving birth to somewhere far out on the living ice. For whitecoats born and raised on the bare open ice are frail, sickly creatures with lacklustre coats and a high mortality rate.

There are those who will argue that our attitude was not a disinterested one, as we would soon go in pursuit of these seal pups and naturally wanted them to be plump and healthy, so that our efforts might be fully rewarded; but to argue thus is to reduce the character of the sealer, a man of the wilds, to the commonest, the most odious of denominators, and to attribute to him the values of the citizen of an industrial, mercantile society. Nothing could be farther from the truth: the true sealer is a friend of the seal!

Take Wellie, for example: in his fifties, six feet tall, with greying hair and high prominent cheek bones that indicate the presence of Indian blood somewhere far back in his ancestry; vigorous, lithe, with gentle, penetrating eyes that are the very image of his character, the very mirror of his soul. He is known and respected as the best sealer in the Magdalen Islands: it is he who provides the curators of the large North American zoos and aquariums with their zoological specimens; and it is he whom the other sealers pursue over the ice during the hunt to be certain of reaching the seal herd, and it is rare that his team does not return with the largest number of pelts. He is a great sealer: he has killed hundreds and hundreds of what the public has come to call "baby" seals – here, we refer to them as *blanchons*, or whitecoats – and yet there are few men who know, understand, and love the seal as he does.

Year after year, he has raised dozens of seal pups in a small enclave beside his stable, awaiting the resumption of navigation in the Gulf to send them off to the various animal-detention centres that have placed orders for them. Some of these young seals have become his children's playmates, following them everywhere, even into the house, where Wellie welcomes them in the kitchen, offering them choice morsels of fresh smelts he catches each day specially for them, as if to reward them for their audacity! He has taken numerous photographs of his children playing with these amphibious creatures, and he shows these snaps proudly to his guests as treasured family souvenirs.

Curiously, he remains attached to those seals he has lodged in his home. Each time he travels to Quebec City, he never fails to pay a visit to "his" seals in the Aquarium, where he seeks out the biologists to whom they have been entrusted, making inquiries about their behaviour, their diet, the state of their health and their spirits; and perhaps he tells himself, as I too have had occasion to do, observing these great adventurers of the sea turning in circles in a tank 15 m × 5 m, that they are less unfortunate when they fall victim to our clubs. But if such thoughts enter his head, he doesn't breathe a word of them.

One day last year, Wellie received a telephone call from the curator of the Quebec Aquarium: the pair of grey seals[2] he had sent there several years before had just given birth to a pup. He was as excited as if he had received word of a birth in the family! If they had reproduced in captivity, they weren't so unhappy, after all. However, these were grey seals, *rouards* as they're called here, a more or less sedentary species that favours the brackish waters of river mouths and the big salmon that harbour there in the spawning season. The harp seal, on the other hand, has never been known to breed in captivity.

Wellie and I had now reached the hummocks marking the end of the shore ice. A channel about two miles wide separated us from the living ice, immobile for the moment in the slack current. Soon, the tide would begin to ebb and would drag this entire ice-field southward at a speed of six knots.

"You've gotta be sure to get back before the tide starts to ebb," said Wellie. "Otherwise, you'll have to spend the night on the ice." He had scarcely uttered these words when the little cakes of ice floating to and fro in the channel began to gather momentum and to clear out at great speed, seized by the invisible force of the current. How right he was: in a small *canot* filled to the gunwales with pelts, with nothing but a paddle to propel you through the icy water, probably with the wind up, you'd be foolish to sit around waiting for miracles.

It's not that spending a night on the ice is such a frightful experience. Indeed, if the weather is fine and there are plenty of seals in the vicinity, it can actually be unforgettable: you erect a makeshift shelter with the *canot*, snow, and blocks of ice; and, with a little fire of scrounged wood – slivers chipped from the paddles, clubs, or the seats of the *canot* will serve in a pinch – you burn the fat of a seal and have heat for the night. If you're hungry, there's no lack of fresh meat. All in all, you sleep almost as comfortably as at home, wrapped in furs, with a live whitecoat as a pillow. If by chance you should freeze your hands or feet, all you have to do is kill a seal, slice it open, and plunge your fingers or toes into its steaming entrails. As long as there are seals in the vicinity, there's no cause for alarm – any sealer will tell you that.

It's the family back home who suffers most, waiting, worrying, passing the night without sleep, while the object of its anxiety is peacefully snoring on his bed of ice, confident that the following morning the tide will have carried him right back to the spot where he was caught lingering the previous evening.

But how can one help but imagine the worst at the smallest delay, the slightest hitch? How avoid seeing oneself as a widow or orphan, alone and heart-broken, when all the folklore of the archipelago is filled with tales of catastrophes and tragedies befalling the men on the ice?

Like my thoughts, the ice in the distance had begun to drift slowly away in the twilight. I tried to imagine the feeling of solitude, alone with three or four other men on a sheet of ice as night began to fall, dull, cold, moving away from the land at a speed of six knots. For while the ice can provide a spectacle of breathtaking beauty when the sun's rays hit it and explode in light – when, rebounding in every direction, the light flows and bends through the translucent prisms of the hummocks and ice statues, distilling fantastic arpeggios of mellow, aqueous colours that run the entire gamut of blues and greens; or when, at sunset, it blazes in crests and waves of glorious reds, streaking the mauve ice-pack with long jagged shadows – it can also be sombre, dismal, even menacing, when the weather is dull, beneath low grey skies that seem to suck out all colour and sound, when from one point to another in the distance all you can hear are the sharp, clear explosions of colliding floes.

It's a world where a man invariably feels humble and small, a world that demands he be constantly in peak condition, for even at the best of times it is unpredictable and fraught with peril.

This is something I learned the hard way a few years ago, one early March morning – March 9, to be exact (the date will remain forever engraved in my memory). It was one of those exceptionally cold winter mornings when the mercury had dropped several degrees below zero on the old Fahrenheit scale, but in actual fact it was much colder because of a strong wind blowing out of the north at sixty kilometres per hour.

That year, the great herd had run smack into the Islands. The seals were everywhere. It would have been a dishonour for any Magdalener worthy of the name not to join in the hunt, even if only on principle. So it was that I found myself far out at sea that morning, in the company of two hearty young lads as anxious as I to prove themselves as Magdaleners on the ice.

To tell the truth, we should have been suspicious of our expedition right from the start. It had been planned only the night before in the heady atmosphere of the Epave Bar, its anticipated success toasted again and again until closing time, after which we had spent the remainder of the night tracking down the various bits of equipment that, needless to say, we were lacking. Our gear, to put it mildly, was makeshift. All three of us were neophytes on the ice. My experience amounted to one sealing expedition the previous year as a reporter for *Maclean's*, and that of my companions was scarcely more extensive. To make matters worse, we had selected as our hunting grounds the worst possible ice, an area at the far end of the North Dune where the great glacier constantly breaks up in the tidal currents and the swift waters of the Laurentian Channel, sending huge fragments of ice swirling in the eddies and whirlpools characteristic of those turbulent waters.

The Magdalen Islands are a natural dyke lying on the axis of the tides. Because of the restraining effect they exercise upon the waters, there are semi-diurnal tides on the side facing the Cabot Strait and the ocean, and diurnal tides on the opposite side: on the south and the east, two high tides and two low tides each day; on the north and the west, only one of each. At either extremity of the archipelago, the conflicting tides meet. The waters held back by the dyke try to escape by circumventing it and run slap into the next rising tide, causing eddies, whirlpools, and undertows that render navigation in the area extremely perilous, and the work of the sealers even more so.

Yes, we ought to have been suspicious of our expedition! Seeing us take to the ice with our makeshift equipment, manoeuvring clumsily, our faces drained of all colour from our excessive libations of the night before, the other sealers, in teams of four or five and equipped with boats and gear carefully assembled a month or more in advance, must have shuddered at the outlandish spectacle. And well they might.

We had scarcely covered the first mile or so, zigzagging through the hummocks of ice and whipped by the bleak, treacherous winds, when *plop*! into the water I fell.

Deciding to take a short cut, I had ventured out onto a pan of ice that had seemed solid enough until I reached the middle of it. There, it began to part company with me, slowly buckling beneath my weight. I have to admit that it served me right for behaving so rashly, and no doubt my two companions were of a like mind, for they stood watching me from the safety of their own ice-floe with looks of utter disbelief on their faces. Up to my neck in water, I found small comfort in the knowledge that I was warmer down there than they were in the open air.

All the same, I had to get out. But this proved much more difficult than it sounds: each time I braced my elbows on the ice to lift myself clear of the water, it sank beneath my weight, like soft ice cream, and all my efforts seemed only to enlarge the pool in which I was trapped and from which I struggled desperately to escape. Meanwhile, the water having soaked through my clothing, which hung cold and heavy on my body, each attempt to save myself seemed to cost me twice the effort of the previous one, with little more to show in the way of progress.

It quickly became clear that I wasn't going to be able to extricate myself from my predicament alone. Strangely, I wasn't particularly alarmed at this fact, only a little indifferent. What I felt was probably something like the fatalism or resignation that men of the sea experience in moments of great peril, that permits them to anticipate the end with the same equanimity they bring to bear upon the events of their everyday lives. I didn't have to make a conscious effort to overcome my apathy, however, for my body was instinctively struggling for dear life to remain afloat.

Not for a moment did it relax. Each time it dropped back into the water, amongst the bits of crumbled ice that had given way beneath it, it started up again, straining arms, shoulders, legs in a concerted effort to pull itself onto the soggy ice that continued to disintegrate beneath it. Yet, little by little, it brought me closer to the solid ice upon which my two companions remained frozen like marble statues, their faces comically resembling masks in a Greek tragedy. Finally, one had the presence of mind to throw me the end of the rope he carried wound about his waist, and I was hauled up onto the ice.

There, the wind took charge of me. And it was petrifyingly cold that morning.

The water didn't even have a chance to drip from my wet clothing: two or three sharp gusts of wind had it transformed into a block of solid ice. As soon as my instincts warned me that I must move at once in order not to freeze to death on the spot, I had to strain biceps and calves to crack my suit of armour at the joints!

Despite its stiffness and its weight – imagine a heavy army-surplus fur overcoat and several layers of woollen sweaters and pants, all saturated with water and frozen stiff – this icy armour protected me from the bite of the wind. Thanks to the heat generated by my body as I walked and the wool fabric of my garments ("Wool will save a man, nylon will kill him," say the old folks), there was a warm, damp zone between my skin and my shell of ice, rather like the spongy space in a diver's wetsuit, and I didn't suffer unduly from the cold during our long trek back to land.

For we were heading back. My singular baptism in the ice had made my day and somewhat cooled the hunting ardour of my two companions. Of course, seasoned sealers don't return home over such a trifle. When a member of a hunting party breaks through the ice – or "passes through," as they say here – he is pulled from the water, stripped naked on the ice and, while one member of the team wrings out his wet clothing, the others slap him with the palms of their hands to keep his blood circulating. Then he climbs back into his clothing and they go on their way as if nothing had happened. But we were heading back, and it was a good thing, too, for my heavy diving suit could have been the death of me at any moment: all I had to do was stop walking.

It was in the pick-up truck that sped towards the first house in Pointe-au-Loup that I really began to feel the cold. Sitting still in the cab, I was no longer immune to the cold; though I tensed all the muscles of my body, I could feel it irresistibly gaining ground. Fortunately, fifteen miles is no great distance to cover in a fast-moving vehicle over a track of hard-packed snow. Then, a roaring fire, dry clothes, hot coffee, flapjacks, and several stiff shots of cognac at the home of the Good Samaritan who took us in – and we cheerfully recounted the details of my misadventure while his wife bustled about, wringing out my wet clothing and hanging it up to dry.

Taking leave of these fine people, I still felt a little ashamed of my early morning blunder. But I knew that I could hold my head high, for almost all sealers have "wet their feet" at one time or another in the course of their apprenticeship on the ice.

This is due largely to the extraordinary diversity of ice conditions. While our western languages contain only a few words to identify its different states (when you've said *ice*, *sleet*, *hail*, and *frost*, you seem to have exhausted the subject), the Eskimo languages – Inuit, Lapp, and Chukchi – each have nearly fifty such expressions, and at least that many more to identify the various types of snow. From experience and observation, a sealer comes to know first-hand several dozen varieties of ice and, although he may lack the vocabulary to name them, he is able to distinguish one from the other by their colour, texture, lustre, and composition, and to evaluate the thickness and solidity of each at a single glance.

They are true scientists, these reapers of the ice-floes! Meteorologists, oceanographers, biologists – they must be a little of each to successfully draw their livelihood from the marine world they inhabit. But more than knowledgeable they are wise, for they do not use their knowledge to dominate nature, to despoil it or to turn it into fodder for machines. Instead, they adapt themselves to it, establishing a direct, spontaneous relationship with it by moulding themselves to it in accordance with the seasons and the slow cycle of time, so that nothing will be greatly disturbed by their passage through that world. There is still a great deal of the "savage" in these men, a heavy reliance upon instinct and intuition, a deep and abiding respect for the wonders of life.

And I gazed at Wellie, at his strong sachem-like profile etched sharply against the greyness of the landscape, as he hopped along in the tracks we had left earlier that day in the snow.

We were heading back to shore. It was all very well to dream of the ice and to imagine ourselves already out in the heart of the seal herd, but night was falling fast and Wellie's day was far from over. Back home he had to milk the cow, feed the chickens, fill the pig troughs, separate the milk. For, like many Magdaleners – and not only the fishermen, although traditions tend to disappear quickly among those engaged in technocratic occupations – Wellie's life is modelled on the old Acadian traditions, in which each family is fundamentally self-sufficient. Each summer he plants a little garden with vegetables, turnips, and potatoes; he raises a number of animals for slaughter and cures the meat himself; he keeps geese, ducks, hens; he catches his own fish and hunts his own game. The only things he is obliged to buy are flour, sugar, and spices.

Clearly, this near self-sufficiency requires discipline and hard work. It is a life that is often looked upon by those not actually engaged in it as a kind of slavery. But Wellie sees it, rather, as an expression of his freedom. The real slaves, in his eyes, are those who live at the mercy of the big food chains and the caprices of the marketplace economy, those whose stomachs in a time of crisis can fall victim to the most contemptible blackmail. As for him, all the currencies in the world could collapse overnight, and it would have very little effect upon his existence on his little patch of earth with its stretch of jagged shoreline.

His chores done, he will return to the house, remove his boots on the porch and probably take a moment to whet his knife on the grindstone, for it must be razor sharp when he comes to sculp the seals. He will toss his catch of smelt into the kitchen sink and pull up a chair at the table, where his wife will set before him a steaming plateful of boiled meat. He will eat slowly, wiping the gravy from his plate with slices of homemade bread, content in the knowledge that his family's comfort is due to his own hard work.

Then, it will be evening. The television set will be switched on in the living room, disgorging its flow of stylized images. But Wellie will only half watch it, for he will be busy tying hooks to his cod lines or mending his herring nets, in order to be ready, once the ice has broken up and carried away the great herd of pagophiles, to profit from the other treasures that the sea, his benefactress, will then offer him.

II
Nazaire

The previous evening, a flock of gulls was spotted in the west.

"They're over there!" Grand Jean exclaimed. And old Nazaire and the others slowly nodded their heads.

Several days had passed since our excursion out on the shore ice, during which time sleet had fallen, then snow, accompanied by violent winds blowing steadily day and night; and that morning, when the Department of Fisheries' helicopter was finally able to make a reconnaissance flight, it spotted a large offshoot of the seal herd not more than ten miles west of the archipelago. The news spread like wildfire throughout the townships, and everywhere the sealers were consulting each other as to the probable course of the ice, bent over marine maps and talking loudly of winds and currents. In the end, there was a general consensus: the seals were headed straight for West Cape!

That same afternoon, Grand Jean, Nazaire, Wellie and several others climbed to the top of a *demoiselle*, a hillock which derives its name from its characteristic womanly shape, and which is so typical of the Magdalen Islands. From that vantage point they spent a long time scrutinizing the distant ice, without detecting the slightest sign of life. Not until they spotted the birds, that is.

"They're heading north," Nazaire said, his hands shielding his eyes from the sun. Once again, the others all nodded their heads. It wasn't necessary to say any more; it was clear that the winds would blow out of the north the following day. No sea-wolf is ignorant of the fact that in the late afternoon the gulls always fly in the direction from which the following day's winds will blow (you can see them passing overhead, their long wings beating the air, as if they are on their way to some prearranged rendezvous); then, during the night, all they have to do is glide effortlessly on the air currents to reach their feeding grounds the following morning at sunrise.

But it was not only the gulls that forecast the winds to be expected the following day; the scarlet sun and the scattered wisps of pinkish-mauve cloud in the western sky confirmed their verdict. Moreover, the radio had predicted fifteen to twenty knot winds out of the north. There seemed little chance they would abate.

It was Nazaire who broke the silence. Spitting out a long stream of tobacco juice, he declared, "The ice'll be loose on the south..." The others nodded, then slowly descended the icy *demoiselle* in single file, not speaking a word, so absorbed was each man in the silent elaboration of his own private sealing plans.

If the whelping ice actually broke up at West Cape, part of the seal herd would pass to the north and part to the south. In the south these floes – or *flardes* as Belgian and French navigators still called them at the turn of the century – would be blown over several miles of open sea, while in the north the wind would push them towards the dunes, piling them up against the ice-pack that could already be seen out there, stretching far out of sight. This would allow for two very different types of sealing: in the north, the hunters would walk out to the seal herd; in the south, they would have to reach it by boat.

Nazaire had long ago settled upon his course of action. He considered himself too old and out of shape to go hunting on foot. Tramping miles and miles over the ice-floe, dragging a heavy *canot* over the hummocks and ridges, then retracing his steps with the *canot* laden with pelts, hauling on the damned rope till his shoulders were raw and bleeding – no, he'd leave that to the younger men. He was a sailor, by God! If the seals were carried to the south, he'd reach them by boat. If not, they'd have to come devilishly close to the shore before he'd go after them on foot.

This he had made very clear to Grand Jean when he had proudly accepted the honour of serving as leader of his team. He had no intention of punishing himself out there on the ice-floes, especially since he knew all too well that there were invariably a couple of shirkers in every team who were content to hold the rope just tight enough not to arouse suspicion, while pulling on it as little as possible, nonchalantly allowing their more diligent team mates to strain like horses. Obviously, he wanted no part of that. Besides, he didn't like the conventional *canot*, which he considered too heavy because of its iron runners.

The craft he had chosen, however, was not one that a polar expedition would ever covet. It was a small plywood skiff that resembled the old-fashioned gufa or the coracle of the Celts, though more rectangular and less elaborate in design, if such a thing is conceivable – a flat-bottomed dinghy, known locally as a *flat*, used for navigation in the harbours in calm weather and lobster poaching in the lagoons. He had equipped this miniscule skiff with a four-horsepower outboard motor, and he was ready to tell anyone who would listen that it was the best conveyance devised for sealing since the wooden sailing vessels of old. He ignored the snickers of those around him; he was used to being looked upon as an odd-ball.

Old Nazaire is a colourful character: Adventurer, Bohemian, rogue, given to ribald pleasures and something of a slave to the bottle, he spent his youth working the waters from the Antilles to Baffin Bay on vessels of the merchant marine; then, for several years he lived as a "savage" with the Naskapi Indians in Labrador. After the War, he married and turned to fishing the high seas. But, one by one, the commercial species of fish – haddock, herring, cod, plaice – began to disappear from the fishing grounds, which were pillaged by foreign fishing fleets, and Nazaire found it more lucrative to turn to poaching and small-scale contraband to support his large family, to which he faithfully added one more mouth to feed whenever he returned from a long time at sea.

Now approaching sixty, radiating an almost magical charm from the fact that he has never been caught by the authorities and has always emerged unscathed from his countless brushes with death, he can often be heard telling the young people who look up to him that it is only outside the law that the truly honest men can be found, and he makes their hair stand on end with tales of wild adventures taken straight from his life.

Like the time he went to Chéticamp, Nova Scotia, in search of alcohol during the Quebec Liquor Commission strike. It was mid-October, and the thirsty population of Havre-Aubert had entrusted him with hundreds of dollars to be converted into precious liquids. He had set out alone on the nearly sixty-mile crossing over high seas in an old lobster boat with a chugging motor that had been hastily put in running order for the occasion.

When he reached Chéticamp, he began to purchase alcohol like it was going out of style, calling upon his friends – the people of

Chéticamp are Acadians, too – for help so as not to arouse the suspicions of the Royal Canadian Mounted Police. But all his efforts at dissimulation were in vain, for the holes that quickly appeared on the shelves of the Liquor Control Board and the activity around his moored boat made it all too clear that something out of the ordinary was happening. A stranger never fails to draw attention to himself in a small fishing village where everyone knows everyone else, and this is particularly true when the stranger happens to be someone like the truculent Nazaire.

But the pirate had smelled a rat. Once his order was filled and he had paid his respects to a local beauty, a lady of easy virtue he promised himself to look up on his next visit to town, he didn't waste a second in casting off his moorings.

It was early morning. There was a strong swell in the harbour, auguring badly for conditions on the open sea, and the wind sent sheets of water spraying over the jetty. The few fishermen who came down to the quay to sniff the morning air called to him: "*Débauche!*[3] *Débauche*! No one goes out today! There's a storm coming up! *Débauche!*" – with the menacing air of strikers trying to intercept a scab. But deafened by the sound of his motor and the whistling of the wind, Nazaire couldn't make out what they were saying, except the one word *débauche*, which in his confusion he took to be a reference to his previous night's behaviour. So he continued to wave at them, a big grin on his face, wondering why they looked so angry, while his skiff began to rise and fall in the billows off Pointe Enragée.

By early afternoon, he found himself in dire straits, beneath a heavy, dark sky that rained torrents on him like a giant waterspout. Massive swells of grey seawater rose on all sides of his boat, then came rolling down on him with a loud, sinister hiss. His skiff was filled with water, his motor was dead, and he had to bail desperately to keep himself afloat, while the bottles, freed of their soggy cardboard cartons, rolled about in the bottom of the boat, smashing to pieces.

All night he pumped and bailed, soaked to the skin, straining to the point of exhaustion, like the mythical Sisyphus: each time he managed to bail out the boat, another wave washed over it and he had to start all over again! And so he toiled, all that day and the following night, pumping and bailing without respite, panting, gasping, chilled to the bone, keeping himself going only by sheer willpower and the salt spray that constantly lashed his face. The following morning, the storm abated.

During the entire time, the wind hadn't ceased to shift, blowing first from the south, then from the east, then from the north, stabilizing finally in a gentle nor'wester; and with no other instrument of navigation than a magnetic compass, Nazaire had only a very vague notion of where he was in the vast Gulf. But he had other things to contend with at the moment: he must bail the boat one last time, sweep up the bits of broken glass, stow what remained of the cargo in the forecastle, and then get some sleep. Later he would try to get the motor working again, but now all he wanted to do was sleep, sleep.

It was nearly a week since the rascal had set out from Havre-Aubert, and the townspeople were beginning to worry. They had telephoned Chéticamp and had been given the details of his latest exploit: "Goin' out there in a little open lobster boat no more'n twenty-eight feet long with a wind of forty knots blowin' out of the south is suicide, nothin' but suicide!" said the older fishermen. But, for all that, they didn't abandon hope of finding him. They tried to estimate the direction and distance his boat would have been carried during the storm and they sent out two boats in search of him; but they scoured the sea in vain. The air patrols sent out by Search and Rescue in Halifax met with no greater success: they swept the skies of the Gulf for several days without finding the slightest trace of the missing man. They were about to call off the search when one fine morning . . .

But here the versions differ.

If you ask Nazaire, he'll tell you that he spent several days fiddling with the motor of his boat before getting it to work (the problem seemed to be with the spark plugs) and then followed a course straight east until he reached coastal waters, from where it was mere child's play for an experienced navigator like himself to regain home port.

Others – whom I'm inclined to consider evil tongues – contend that there is not a scrap of truth in his story.

According to them, Nazaire tried unsuccessfully to repair his motor and in despair seized the first bottle he could lay his hands on to comfort himself in his plight. This he followed with a second bottle, and then another, thirsting as only a sailor in distress can thirst; until, the days succeeding one another as the brandy succeeded the vodka and the rum and the gin, he became royally drunk, drunker than he had ever been in his entire drunkard's life, a fact to which the cod fishermen who had picked him up as he drifted aimlessly not far from the Georges Bank, were ready to testify.

When they sighted the boat, there was no more sign of life on board than on the Flying Dutchman, so they approached with extreme caution, expecting at any moment to see it transformed into St. Elmo's fire or some other nightmarish marine apparition. They circled the vessel, keeping their distance, calling their comrade's name at the top of their lungs, without receiving any response. Then plucking up their courage, they prepared to board the wreck, tears in their eyes, fearing the worst.

And what did they find, wallowing in the bottom of the hull, dead drunk but still conscious, a half-empty bottle in one hand and a dazed smile on his face? Their old friend Nazaire, of course! He had drunk nearly all the alcohol he had been entrusted to deliver to the good folks back at Havre-Aubert. And the scoundrel never did reimburse them for their loss.

And it was this character that Grand Jean, Arthur, Camille, and I had chosen to be leader of our sealing party that spring. The citizens of Havre-Aubert could snicker to their hearts' content, however, for there was method to our madness. The old pirate took his role to heart, as if our own enthusiasm had inspired him with a new spirit of youthfulness and responsibility that surpassed even our own expectations.

From the moment Grand Jean approached him with his proposal, he began to moderate drastically his consumption of alcohol and threw himself body and soul into preparations for the expedition, rigging out his old *flat* for the ice, braiding rope ends, hewing clubs, shaping hooks, splicing hawsers, tending to the most minute details of the equipment, and lecturing us at great length about sealing. Under his tutelage we were all ready for the great day when he came, in the early hours of the morning, to awaken us.

In no time at all, we slid into our clothes and ate our breakfast, then emerged into the sharp, crystal-clear night, beneath a sky literally strewn with stars. Those first few moments of our excursion in the silent open stretches of the Gulf were almost magical, the icy snow squeaking beneath our rubber boots, the vapour streaming from our nostrils, the long frost-stiffened ropes gripped in our fists, an inner jubilation filling our hearts. Together, we lifted the *canot* onto the trailer behind the truck. Far to the east, we could vaguely make out the first pale rays of the dawn.

A number of other sealers were already out on the shore ice when we arrived, pacing to and fro, sniffing the air. It would be a fine day; it was written in the sky. And yet they hesitated to set out; they wanted to be certain of the wind. Some even preferred to wait until sunrise before committing themselves.

In this way there are always some (and surely they are the real "landsmen") who spend the entire day perched on the cliffs, scanning the distant ice-floes until sunset, trying to follow the movements of the other teams in the maze of hummocks and leads, and spreading the most absurd rumours about the fate of those momentarily lost to sight. Thanks to these expert analysts of the sealing scene and to the zeal of other folk concerned to quickly spread the news, Grand Jean's parents received no less than seventeen telephone calls that day announcing our deaths somewhere out there in the icy Gulf!

But Nazaire was not one to hesitate on the threshold of adventure. After briefly consulting with the other leaders, he signalled to us to lift the *canot* off the trailer. Then he sauntered up to us, winking. "If we hang around here, boys," he said, "we'll just get cold feet!" He passed the lead hawser over one shoulder and set a quick pace for our short trek across the shore ice.

Then things began to move: the *canot* was in the water, we were in the *canot*, and the four-horsepower motor sputtered and caught, breaking the silence with its quiet purr.

Seated in the rear of the skiff, I gazed at the black water swirling away behind us and the little waves breaking against the bluish ice of the shore. Our example had shamed some of the other teams into emulating us: they were taking to the water, preparing to give us chase, paddling like mad, in the best tradition of inshore sealing. At the helm, Nazaire went on giving us final instructions, like the coach of a hockey team before an important game. "When we get to the seals, don't hold back, boys, kill all you can lay your hands on, kill all the seals you see. There's no time for loafing out on the ice."

He outlined his plan for the hunt: on the far side of the two-mile-wide channel we were now crossing, there should be several patches of seals, perhaps even a fair-sized offshoot of the great herd. The problem was locating it. For the moment, since the tide was rising, shoving the pans against each other and forming a compact floe, we would follow a straight course for the far shore, from where we would proceed on foot. Later, when the tide began to ebb, opening leads between the floes, we would be able to put the boat back in the water.

Dawn was breaking as we approached the living ice. Suddenly, we heard a heavy throbbing sound above our heads, punctuated with sharp, staccato cries.

"Crows!" exclaimed Nazaire. "They're headed for the ice!"

There were hundreds of them, flying in close ranks, their heavy wings beating the air, moving in a direction only a few degrees west of our course.

"That's it," he said, "the seals are over there," and he gave a light tap to the tiller, turning the craft slightly to starboard.

We were flabbergasted, unable to credit the amazing sight we were witnessing, much less the stupefying conclusion drawn from it by our skipper. An explanation was in order.

"When the females whelp," said Nazaire, "they leave all kinds of odds and ends on the ice . . . membranes, umbilical cords, placenta, sometimes even still-born pups. So if the herd doesn't stray far from shore, the crows get wind of this appetizing feast and leave their dumps and garbage pails for a while to go and feed on the ice."

We drew alongside the ice-floe without a hitch; then, with the hawser over our shoulders, we began to move swiftly in the direction taken by the crows, hauling the light skiff over hummocks, fields of solid ice, and rock-hard billows of snow. The dull light of dawn lent the icescape an eerie, lunar atmosphere. A cool wind blew over us, and the immense silence of the Gulf was hardly broken by the creaking of our cleats biting into the ice, the squeaking of the boat, and the deep rhythmical sounds of our breathing.

Suddenly, Nazaire signalled a stop. We stood, listening. In the distance, we could just make out a diffuse murmur, a vague incessant humming that seemed to rise out of the landscape itself. The old man smiled. Then, almost at once, he resumed walking at an even swifter pace.

Little by little, the music increased in volume. It was a plaintive sound, at first soft with strangely pastoral tones, like a lament woven with tears, wails, whines, yelps, all merging together into one long, endless complaint – the incredible chorale of the whitecoats! There, straight ahead of us, tens of thousands of seal pups, spread out over several square miles, were simultaneously uttering their distinctive call. (It is by means of this call, it is said, that the mother seal is able to locate her pup in the herd.) Their strange concert of lamentations was gradually transformed, the closer we approached, to a throbbing, discordant refrain, rising finally to a great clamour of savage, deafening cries that, though painful to the ear, possessed a grand, majestic power, like the roar of the cataract at Niagara.

Curiously, this din very quickly got on my nerves. I felt deep within myself an intense resentment, an acute irritation, a fierce desire to silence this tumult of caterwauling and restore silence to the scene. "Poor little whitecoat!" I thought. "You think your colour camouflages you on the ice, but you give yourself away with your shrill cries, your incessant complaints! And it's the sound of those cries that unleashes this murderous hatred in me, this secret desire to kill you."

Of course, this was easier said than done.

We arrived at the summit of a crest overlooking a basin strewn with stunted hummocks. From small waterways criss-crossing the ice there appeared and disappeared at regular intervals, with the rhythm of some strange ballet, a number of shiny, black, cowled heads – the harp seals! There must have been several dozen of them performing this unrestrained saraband. Nearby, under the hummocks, lay other seals, seemingly oblivious to all this revelry. These were the mothers; they lay beside their pups, tufts of silken hair with big round black eyes, most of them in groups of two or three, some alone, dozens of them spread out over the ice, perfectly immobile. There was in the cast of their features an air of infinite indulgence for the exasperating cries of their noisy offspring.

The beauty of this spectacle, this sudden intense abundance of life, took us completely by surprise, and we stood for a moment without speaking, feasting our eyes on the amazing tableau from the height of our promontory, stunned, entranced, unable to move. Then, alerted no doubt by our odour, several of the mothers began to waddle toward the openings in the ice and to dive into them, at the very moment that Camille and Arthur, followed closely by Grand Jean, ran swiftly into the middle of the patch, uttering Indian war cries and swinging their clubs wildly over their heads, sowing panic in the herd. On all sides, the mother seals threw themselves into the water, abandoning their pups on the floe.

Nazaire glanced at me, and I understood that I must join the others. Already, the clubs of my companions were rising and falling. The kill had begun. Somewhat reluctantly, I took off down the slope.

"Here! I'll show you how it's done!" said the old man, taking me by the arm and leading me up to a pair of whitecoats that were huddled on the ice, wailing pitifully. He seized one by the fur of the neck and set it before him, then raised his club and dealt it a heavy blow to the skull. There was a dull thud. The body of the whitecoat stiffened, as if in protest against the violence of the impact; then he struck it again, this time on the brow, causing the blood to gush from its nostrils and eyes.

The little creature was shaken by a few spasmodic convulsions, then lay still. It was dead.

Without a moment's hesitation, Nazaire rolled it onto its back and opened it from chin to tail with a single, quick slash of his knife, whetted as sharp as a scalpel. Then he thrust this formidable weapon into the animal's heart and quickly rolled the body back onto its belly to allow the blood, which was gushing from the corpse in big red bubbles, to escape without staining the fur. The entire operation, which lasted no more than a few seconds, resembled a sacrificial rite. But I had no time to contemplate this, for he suddenly said, "Your turn," seized the other whitecoat, and deposited it at my feet.

I felt a shudder run down my spine as I gripped my club, and the first blow I dealt must have reflected the intense conflict raging within me – to be or not to be a sealer – for it was a light blow, without conviction, and to my great shame it struck the ice with a sharp crack several inches off target. Nazaire gave me a knowing look. "You got to get yourself worked up to club them," he said. "It's no good if you don't. You can't do that sort of thing casually. Think of something that gets your back up, something that makes your blood boil. Listen to them. Have you ever heard a more irritating sound? That's how the rest of us work ourselves up for the kill, we think of that.''

I gritted my teeth and once again raised my club. This time I struck a direct blow to the skull of the pup. I struck it again and again and again – a shower of blows that Nazaire finally had to stop. Then, somewhat to my astonishment, my right hand reached instinctively for my knife, as if this gesture had always been a part of my bodily reflexes. Taking the pup by the flippers, I rolled it onto its back, as I'd seen Nazaire do a few moments before, and sliced it open from top to bottom. Without a moment's hesitation, I plunged my knife into its heart.

Then Nazaire knelt beside the little corpse with an air of intense, almost solemn concentration, a gesture that seemed as out of keeping with his piratical appearance as a blackhead on a starlet's nose. Cupping his hands, he scooped the blood from the cavity of the animal's chest, lifted it straight to his mouth and drank it in a single gulp.

I had heard how the Montagnais Indians of Ungava and Labrador drank the blood of the caribou they killed, lapping it straight from the gushing artery, to replenish their strength before carting the meat back to camp; that was in the days when hunters still used bows and arrows and had to run miles and miles through the forest in pursuit of a wounded animal before bringing it down. I was not unduly surprised, therefore, to see Nazaire, who had lived for some time with the Indians and whom I knew to be an unusual character, perform this strange ritual.

What bewildered me, however, was to see him smear his cheek bones and eyelids with the blood that remained on his hands, carefully, expertly, like a woman applying make-up.

"The morning sun is bad for the eyes," he said, in response to the bewildered look on my face. "It can cause snow-blindness."

But seeing me still perplexed, as if I had been struck dumb by the events I had just witnessed or dazzled by the scarlet sun that was now setting the ice-floes ablaze, he gave me a more thorough explanation.

His father, "the old man back home," had taught him the technique, he said, the first time he took him out sealing when he was still a boy. He had learned the practice from *his* father, who had learned it from *his* father, and so on right back to the Mic-Mac ancestor who almost invariably figures somewhere in the genealogy of most Acadian families. You must drink a mouthful of the blood of the first seal killed each day in order to keep up your strength; but never more than a mouthful, because if you absorb too much seal blood, the effects of its high concentration of vitamins and mercury are apt to be debilitating rather than stimulating. Then you must smear some of the blood on your eyelids to diminish the blinding reflection of the sun and thus protect yourself against snowblindness.

"But for the Indians," he went on after a short pause, "there was more to it than just its practical benefits. And that's true for us, too, even if we won't admit it. It's a sacred act, a religious ceremony. It's a communion, you might say, a marriage: when I drink the blood of the seal, he comes to life inside me and he becomes my brother. That's a common belief amongst the Indians. Then the Great Spirit of the seals understands that I don't attack my brothers out of hatred or cowardice, but from necessity, because we're all equals out here on the ice. And because the Spirit of the seals understands from my act that I'm his blood brother, and the brother of all nature, he gives me the strength to continue the hunt."

I was sure the old man was exaggerating, but I also had the distinct impression that there was a grain of truth in his account. So, I had to come all this way to understand! To think that all the television programs and magazine articles that decry the seal hunt and denounce the "massacre" of the "baby" seals, have used the very blood with which the local hunters smear their faces as an illustration of their "bloodthirsty cruelty" and to present them as "sadistic torturers," while, at worst, it is no more than a defensive measure against the blinding light of the sun and, at best, a symbolic expression of the union of predator and prey.

But, once again, the old man didn't give me the time to reflect upon these thoughts at any length.

"Your turn," he said, indicating the little hairy body that lay at my feet, the geyser of blood already beginning to abate.

There was an ardent, passionate, magnetic gleam in his eyes, which looked almost hallucinatory in the scarlet mask of his face. There was no time for hesitation. Despite the disgust I felt at the thought of drinking fresh blood I dropped to my knees, trying to shut off all my senses, scooped up a handful and quickly gulped it down. Almost at once, far from the nausea I had expected, an invigorating warmth – almost like that following the consumption of raw alcohol – pervaded my body, spreading through my limbs at lightning speed. And the blood left no disagreeable after-taste in my mouth.

The union had been sanctified.

Nazaire looked at me and winked. I smeared my fingers over my eyelids and cheeks, as I had seen him do. And for the first time in my life, I felt that I belonged out there in the kingdom of ice.

III
The First Hunt

The sun was blazing down on the ice-floe.

We had left the spot where we had killed our first whitecoats, about twenty in number, and were making our way towards another patch of seals, another offshoot of the great herd that we had spotted in the distance from our vantage point atop an icy pinnacle. Other teams, including those that had followed us out onto the ice, were working the seals in the immediate vicinity. Weighed down with pelts, the *canot* was much heavier now than it had been at the outset, and it was with great difficulty that we manoeuvred it through the gorges and passes of the large hummocks and steep ranges we had to cross. "The ice has sails," said Nazaire, referring to several large sheets of ice that rose straight into the air, several times the height of a man, and that roughly resembled the sails of a felucca. Certainly they functioned as such. The tide was ebbing and we were drifting fast in a southerly direction, beneath the double impetus of current and wind, which had not abated significantly with the arrival of daylight.

Being apprentices, it had taken us nearly an hour to sculp the whitecoats. In keeping with his traditional role of leader, Nazaire had skinned only a couple of the beasts himself, and these only to demonstrate the technique. He explained that the master watch or leader of a hunting party should never participate in the actual hunt, for he might become too involved in what he is doing and forget to keep an eye on the *canot* or attend to the movements of the ice and the changes in wind and temperature – "at the risk of losing his men," he added. He was content, therefore, simply to supervise us, moving from one to another, offering comments, criticism, advice.

Removing the skin of a seal is a delicate operation. You can't tear it off as you would a hare's, pulling, panting, cursing, as some self-styled witnesses would have it. With all due respect to those who rise up in indignation at the image of poor little "baby" seals being skinned alive by a gang of bloody brutes, that is not at all the way it is done.

Working from the initial incision that runs the length of the animal's body, you carefully carve away the sculp – the fur and the three or four centimetres of fat adhering to it – by slicing as close to the carcass as possible with the curved blade of the sculping knife, using it like a scalpel. Wielded properly, it will slide between the flesh and the fat as gently as through soft butter. You must be careful, however, not to nick the hide or to touch a bone (in which case you'll have to stop and regrind your blade) and, above all, not to cut yourself, for the seals are carriers of a serious infectious disease, known locally as "seal finger," that can cause inflammation in a wounded finger. Such a wound may even result in the loss of the finger if the sealer neglects to have it treated at once, for the days of the seal hunt are followed swiftly by the herring, lobster, and cod seasons, and a wound plunged repeatedly into water, oil, salt and tripe, will not heal and the swelling will persist – to the point that, one day, unable to bear the festering finger any longer, the hunter will lay it on a stump and sever it with a blow of an axe.

The seal pelt is separated from the flesh by sliding the knife in concentric, semi-elliptical circles, moving from the head to the tail, now to the left, now to the right. Heavy with fat, the pelt pulls away from the body, as long as the sealer remembers to turn the blade slightly with each stroke to conform to the oval shape of the body, as he cuts deeper and deeper into the wedge.

In the vicinity of the flippers, there is an obstacle where most neophytes take the edge off their blade: the clavicle. Because it's hidden beneath fat and meat, you come upon it unexpectedly and can hit it hard; then you have to stop and resharpen your blade. With experience, you come to know the anatomy of the seal by heart and you learn to slip your knife through the narrow space separating the clavicle from the humerus; but in the beginning it's wiser to work your way around it and approach the shoulder from the rear, sliding your knife along the slope of the shoulder blade and down into the tender cartilage of the joint.

The process is completed by making a hole between two of the ribs, passing a finger through it and lifting the carcass out. The heavy pelt drops from the body, and all that remains to be done is to slice through the distended tissue of the back to separate it from the carcass, moving from the snout and the whiskers down to the wee bit of tail located between the rear flippers.

I won't deny that this operation is almost as difficult to execute at least at the outset, as it is to describe; but like anything else it can be mastered with practice. A good sealer takes scarcely more time to sculp the skin of a whitecoat than the average person would take to read this description of it. The best of them can carry out the entire operation in thirty-five seconds flat, which may account for the impression some people have that they skin the beasts alive. With time, all the movements that constitute this difficult art blend into a single, harmonious, rhythmical act, in which speed is a factor of the utmost importance. Strangely, it's not with the eye but with the ear – the characteristic sound of the blade sliding quickly between the meat and the fat – that the expert sculper guides his knife, knowing from moment to moment whether or not he is on the right track.

Once the pelt is removed, it is laid out on the ice, hair down, to cool. (Laying it hair up would warm the fur and greatly diminish its quality.) When all the animals are sculped, the pelts are piled, fat to fat, fur to fur, in the boat. Then the sealers take off in search of more seals, hauling the *canot* over the ice at a pace always set by the lead man.

Obviously, we were only in the infancy stages of this complex procedure on that particular morning, and amongst the hides we had in tow in the flat, some looked more like Belgian lace than sealskins! The old pirate, Nazaire, didn't stop railing us for our awkwardness. He was obliged to admit, however, that the quality of our work had significantly improved by the third or fourth pelt: the little jerking movements of the knife, as dangerous for the fingers as for the skins, had given way now to long, even strokes, and it was only the flippers, or occasionally the snout or tail, that still gave us difficulty. The pelts, at least, were all in one piece.

We reached the second patch of seals, on another ice-floe, after a good twenty minutes' walk. This one was much larger than the first. It covered a floe a few acres in size, broken here and there by little leads that all converged on a vast stretch of dark, rough slob ice. Hundreds of whitecoats were scattered all about this lake of crumbling and disintegrating ice – so many that it was impossible to take them all in at a single glance. In the leads, filled with bobbing discs and slabs of ice, we could see the black-cowled heads of the adult seals rising and blowing; but they were no longer performing the unbridled sensual ballet we had witnessed earlier in the day. Perhaps this was due to the fact that a large number of the females had gone off to rejoin the males – judging, that is, by the number of pups left unattended on the ice. Their cries filled the air with an outlandish din.

"Should be enough here to keep you busy all day," said Nazaire simply, dropping his hawser. He turned and swept his gaze slowly over the area. "And try not to be so clumsy this time . . ."

He had no intention of joining us, the tone of his voice made that clear. He had other things to do. He watched us withdraw, without haste, in the direction of the seals, our clubs over our shoulders; then, when we had set to work, he set off in the opposite direction, skirting the lake of slob ice, and began to scale the iceberg that locked it in on the south. From there, he could survey the entire icescape, determine its drift and study the labyrinth of channels and waterways that the ebbing tide had opened up between the floes. All this knowledge would be valuable to him later on, for soon he would have to turn his mind to the problem of transporting our first load of pelts back to land.

He returned in our direction a short while later, consulting his watch, an anxious look on his face. I saw him speak with Grand Jean, pointing first in the direction of the Millerand lighthouse, which we had passed earlier in the day, then towards the Havre-Aubert buttes whose profiles stood out sharply in the northeast, downstream of our position; and Grand Jean nodded his head vigorously in agreement. I couldn't hear what they were saying but I could guess the gist of their conversation. My suspicions were confirmed a few minutes later when, passing my way, Nazaire told me to begin sculping the beasts I had killed. He wanted to head back to land as soon as possible.

His plan was an astute one. By getting under way at once, he would let the ebb tide carry him to Havre-Aubert dune, where he would unload the pelts; then, still riding the tide, he would reach our ice-floe, which would have drifted meanwhile far downstream of its present location, almost as far as Entry Island. But his strategy didn't end there. By the time he was ready to head back to land with his second load of skins, the tide would have begun to rise again, the current would have changed directions, and once again he would be able to rely upon it to carry him diagonally across the channel to the dune, where he would unload the boat before returning to fetch us.

I have always marvelled at this extraordinary sense of the relativity of things, in matters of space as well as time, enjoyed by men of the sea: their instinctive comprehension of the complex phenomena that control the behaviour of enormous bodies of air and water, in relation to which they are as insignificant as a drop of water in the ocean, and their ability to profit from these phenomena. In my opinion, it is this ability, not to defy nature but to assimilate oneself with it, even in its most hostile forms, in order to draw one's sustenance from it, that sets these men of the sea apart from the mainstream of humanity. In this respect, our friend Nazaire had nothing to learn from anyone: he was the master of every contingency.

Such was the nature of my thoughts as I watched the *flat* withdraw, filled with our first load of pelts, the boat so full that its gunwales seemed almost level with the water.

Arthur had accompanied Nazaire. We could see them zig-zagging down a sinuous lead, now and then pushing aside large blocks of ice with the oars to clear a passage for the boat. Then they passed behind a ridge of high hummocks and disappeared.

During their absence, Grand Jean, Camille, and I continued to work the seals: while two of us sculped the animals, the third panned the chilled pelts, attached them by passing a rope through the eye holes, then dragged them to the loading zone, at least a quarter-mile away. The sun was now high in the sky. Its balmy rays kept the air at ice level comfortably warm, despite the cool breeze that blew over us. To work at ease, we had to remove our woollen sweaters, wearing only our shirts beneath our oilskins. Even at that, we were all sweating profusely.

It's not an easy business making your living on the ice. It is a strenuous, merciless task, with no time to rest, and, as the day wears on, you begin to pray with all your heart that the leader will give the signal to turn back. Clubbing, sculping, towing, heaving, panning, running; all the muscles of your body, stiff from the long inactivity of the winter, protest against this sudden brutal awakening. But the pains in the back, the legs, and the shoulders only serve to intensify the determination, ambition, and proud competitive spirit that drive each team, and each man on each team, to try to outdo the others.

For that is the law of those who live off the sea: each man must do his share, and that share is the most possible. I recall hours, interminable hours, sitting in a stupor on the wet deck of a trawler, slicing and gutting cod, thousands of cod, one after the other, my mind far away, lulled by the rocking of the waves and the monotony of the gestures I had to repeat over and over. The wind might come up, the seas might rise and be crested with foam, the waves might splash over the deck, but I was oblivious to all that as long as there was room for one more fish in the hold!

The same goes for sealing: once you have mastered all the movements that comprise the routine, you work in a daze, oblivious to the drift of the ice-floes, the course of the sun in the firmament, and the sudden storms and blizzards that can abruptly change the face of the day. Nothing matters but this titanic task you have set yourself and that you constantly curse beneath your breath! And then, by scorning your pains and deliberately driving yourself almost beyond the point of endurance, you reach a point at which you no longer feel anything but a sort of euphoria in which everything – the bite of the wind, the acrid odour of seal fat, the red of the blood on the white snow, the deep blue water, the bright yellow oilskins, all the unbridled activity that surrounds you on the ice – merges into one homogeneous, yet chimerical, whole.

So mesmerized were we, that even the return of Nazaire and Arthur at the stroke of noon seemed as strange as the arrival of visitors from another planet. And hadn't they, indeed, come from another world? They spoke of the current, the rising tide, the choppy water of the channel that had given them so much trouble, with the boat so laden with furs. Nazaire thought the wind might drop later in the afternoon, but he wouldn't put too much faith in the possibility.

We quickly ate our meagre lunch of sandwiches and molasses biscuits. Then we loaded the boat with pelts and it departed once again for land.

The afternoon passed like the morning: up to the elbows in blood and fat, repeating the same movements over and over, wielding the club, the knife, the rope. Bit by bit, I was learning my craft. The reticence, the uneasiness, the terror I had experienced clubbing my first whitecoats had given way to a feeling of detachment, a feeling almost of indifference, a total absence of sentiment that made me sense that all the carnage in which I was immersed existed somewhere outside myself. I had lost all trace of compassion for my victims: they had become, for me, inanimate things. This was strange, for I am no hardened killer. I have never hunted any animal other than seals, I have never shot at a living creature, and I still feel a slight squeamishness whenever I have to cut the throat of a chicken or a pigeon for the pot! And yet, there I was, bringing my club down again and again on the skulls of these helpless whitecoats, as if it were the most natural thing in the world.

To tell the truth, the irony of this situation was lost on me on the ice that afternoon, so immersed was I in my work; it was only later that I gave it any thought, and concluded that there was nothing abnormal in my behaviour. I was merely assuming my role as a man, a beast amongst beasts, simply, unhypocritically, subject to the inexorable laws of nature that dictate that one must kill to survive, that there are predators and prey and that the phenomenon we call life passes from one to the other, from the steak to the man who eats it, like the links of an interminable chain.

Nazaire and Arthur returned for us much later than expected, since the lead by which they had planned to thread their way back to our ice-floe had closed up with the rising tide. They had been forced to climb onto the ice at its outer edge and to proceed on foot to our location, dragging the empty *canot* behind them. Now we had to retrace their steps, a distance of perhaps two and a half miles, with a full load of pelts. It was a great effort to demand of our bodies, broken with fatigue after an exhausting day's work, our shoulders sore, our feet so numb we had difficulty putting one in front of the other. Secretly, I thought I would never make it. But there are unexpected energy reserves in the human body: once old Nazaire set the pace, we all felt a sudden rush of adrenalin, and the *canot* started to move swiftly over the ice.

Filled as it was with at least a half-ton of furs, fat, and meat, it was no simple matter getting it across the hummocks and pressure ridges that seemed to multiply at will along the route. We had to pull, push, strain like oxen to move it inch by inch up the steep slope of a mound of ice; then, reaching the top, we had to dig in our heels to hold it back, clinging to it desperately to prevent it from breaking loose and crashing into a sharp spear of ice on the far side. There were stretches of solid ice and hard-packed snow over which it slid like a charm; then there were more obstacles to overcome. We all sighed heavily with relief when, from the top of a hummock, we saw the rough water of the channel glittering in the distance.

Clearly, there was no question of fitting all five of us into the already overloaded *flat*, so we tied about two dozen pelts together with a nylon line and took them in tow. In this way, we managed to reach the shore ice without too much trouble, though we had to bail the boat more than once in the choppy water.

In the distance, the cliffs were speckled with human figures, a dense, motley, restless crowd gathered to welcome the sealers home, as is the tradition on the days of the seal hunt in the Magdalen Islands. There were old men with field glasses slung about their necks, harking back to the days of their youth and spitting long streams of tobacco juice; children chasing each other in and out among the legs of the adults; women, worried and tense until they perceived in the distance the men they were anxiously awaiting; teen-agers trying to impress other teen-agers by making their skidoos backfire; and, finally, the land-locked sealers, the real slackers, who consoled themselves for not having ventured out onto the ice with flasks of hard liquor in the comfort of their automobiles. The atmosphere of festivity could be sensed even from afar, for to the sealer returning from the open sea, the crowd resembled nothing so much as a nation of Lilliputians. To one and all, he was the hero of the hour!

As soon as they thought we were close enough to read their signals, men stationed high on the cliffs sent semaphore messages in our direction, indicating the best spot to land. Then the crowds surged down onto the shore ice. Everyone wanted to be the first to seize the mooring line, to steady the *canot* against the bank, to assist us ashore. There was nothing for us to do but savour the welcome that had been prepared for us by allowing ourselves to be transported ashore by the forests of arms stretched in our direction.

The *flat* was hoisted out of the water by several strapping young lads and made fast to the snowmobile that hauled it to the top of the slope. The crowd separated to let us pass, forming a sort of honour guard for us, and as we passed through it, we were clapped on the back and bombarded with questions: How many? Where? When? What was the ice like? Did we meet so-and-so's team . . .?

Nazaire brought up the rear, answering all questions fired at us, shaking hands, suffering embraces, drinking straight from the bottles passed in his direction, inquiring about the performances of the other teams on the ice. In the midst of all this gaiety, he was jubilant: our team had returned with the greatest number of pelts.

Up above, the old men examined the catch. "You've brought back the meat," they said, handling the bloody carcasses with affection. One, his nose dripping, his cheeks hollow, wrapped in a shabby coat beneath which his thin bones rattled, came up and asked me in an almost inaudible voice if I would give him the liver of one of the animals. I cut it out and dropped it into the wrinkled bread wrapper he held out. He didn't say a word; but as he looked up at me before hobbling away, I saw there were tears in his eyes.

IV
The Art of Sealing

Needless to say, we didn't have to be lulled to sleep that night, and when Nazaire came to awaken us long before daybreak the following morning, we would have been quite content to sleep another hour or two. Oh, the pain of dragging the old body out of bed! My carcass baulked at every movement, stiff as it was from head to toe; and if the old pirate hadn't been there to goad me on with his sarcastic remarks, I sincerely believe I would have forsaken the ice, the seals, and the whole business right there on the spot.

"If you're afraid of a little pain and suffering, you'll never be a sealer," he said as he heated the water for tea. "I'm going to show you good-for-nothings what it's like to suffer!" Grand Jean and Camille bellowed suggestions as to what he could do with his suffering, but the old pirate was in fine form that morning; he had a quick reply for every remark, every sarcasm sent in his direction. We were all in high spirits by the time we had finished our breakfast and hit the trail.

Suddenly, we were in a jesting mood. Rather than carefully descending the slope to the shore ice, guiding the *canot* over the rough terrain as we had done the previous morning, we climbed into it and flew down the hill, shouting loudly as we went, like revellers on a roller coaster at a county fair! The other sealers couldn't believe their eyes. No doubt they took us for fools, and they were not far wrong, for if we had struck the smallest chunk of ice, our vehicle would almost surely have been wrecked.

Since there still remained a large number of whitecoats to harvest on the ice we had worked the day before, it was in that direction that Nazaire once again pointed the prow. The weather had greatly deteriorated during the night: it was still quite mild, but a cool wind had come up out of the west and thick heavy clouds scudded across the sky. The weather forecast called for gusts out of the south-east accompanied by snow flurries, but as yet there were no signs of either. A number of other teams had followed our example in setting out early and were already headed for the ice-floe, straining heavily on the paddles.

As we passed them, Nazaire made as if to throw them a line and take them in tow, to which they replied by shaking their fists, giving us the finger, and firing off a long string of abuse in our direction. But beneath the teasing and the hooting there was something profoundly symbolic in these exchanges. What I was witnessing was the encounter of tradition and novelty, the old and the new. Most of the sealers still clung to the traditions of their ancestors, and they were justly proud of the fact. Our little motor, as insignificant as it was, placed a gulf of two centuries between us. In their eyes, it constituted an unhappy adulteration of the old ways, a crutch, an obscenity. And they were not at all reluctant to make their feelings known to us.

That morning, we were not alone on the ice-floe. News of our success the previous day had made the rounds of the Islands, and several other teams had hurried to join us. There were sealers from almost every township in the archipelago: men from Havre-Aubert with their affected French, others from Fatima and l'Etang-du-Nord with their loud guttural accents, still others from Havre-aux-Maisons with their curious diction deprived of all "r"s; there were even some anglophones from Entry Island yelling to one another from one end of the ice-floe to the other at the top of their lungs to make themselves heard over the cries of the seals.

They were all first-rate sealers, wielding the knife with great authority, and sculping the whitecoats at an astonishing speed. As impressive as their performance was to our untutored eyes, however, it was the men whose job it was to retrieve the pelts and pan them who afforded the most extraordinary spectacle.

Rather than repeatedly circling the leads and streams of slob ice to fetch the pelts and return with them to the *canot*, they had perfected a whole variety of techniques to get themselves across these waterways. A lead would be crossed by leaping from ice pan to ice pan as they drifted by, rather like a man leaping from rock to rock across a mountain stream: copying, they call it. The slob ice, on the other hand, was dealt with in two ways: when his hands were empty, the sealer would propel himself across it at great speed, like a rabbit scampering through mud, pedalling furiously, water half-way up his boots, until he reached the farther shore. On the return trip, he would carry two pelts, placing first one, then the other, in front of his feet to serve as stepping stones, then picking them up behind him, constructing a sort of mobile bridge that stood only long enough to let him cross.

Intrigued by the ingenuity of this technique, Camille and Grand Jean set out to master it. As for myself, I had too many bad memories of slob ice to risk such a feat. But not wanting to appear more clumsy than the rest, I set to work crossing the smaller leads by jumping from ice pan to ice pan, which in itself required a great deal of nerve. The slightest hesitation or loss of balance would have meant another icy bath.

Watching the number of balancing tricks practised by the other sealers, some of whom had reached an age at which they probably no longer found much amusement in such capers, I was astonished not to see a single mishap. Any slip would have been tragic, for like the vast majority of fishermen, most of these men do not know how to swim!

In the old days, the sealer's equipment included a gaff, a long pole with a hook at one end, which he used not only to strike down the seals and to drag the pelts to the pans, but also to catch hold of the ice if he accidentally fell into the water. It was, simply speaking, his life-line.

Today, as a result of the public campaign waged against the "massacre" of "baby" seals – there are those who claim that some hunters actually used the hook to torture the whitecoats! – the rules of the hunt forbid the use of this particular piece of equipment. For humanitarian reasons, it is claimed.

It was late in the afternoon on the second day of the hunt that I witnessed one example of cruelty – the only such incident, I may add, that I have ever observed on the ice. The other teams had left the ice-floe, their *canots* filled with pelts, and we were expecting Nazaire and Arthur to return at any moment to pick us up. Grand Jean, Camille, and I were working a new patch of seals on the far side of the icebergs that flanked the pool of slob ice. As usual, the mother seals had plunged into the water at the first sounds of our approach. All but one, that is, who had decided for some reason to stay and defend her pup.

It was clear at a glance that this pup was a pampered creature.

Fat as a monk, it had thick, silky, dazzling, white hair, which must have whetted Grand Jean's appetite the moment he laid eyes on it.

Not all whitecoats are alike. Some are small, sickly creatures with dull, flat, yellowish hair. These are the orphans, pups who have been abandoned by their mothers or lost during a storm. The majority are plump and hearty, the object of more or less regular maternal affection. But some are simply superb, the spoiled tots of the pagophile herd, pups whose mothers are on hand day and night to respond to their slightest whims.

It was one of these rare specimens that Grand Jean was now trying to abduct from its mother. But each time he came within reach of it, she would rise up on her flippers and utter fierce growls, rush him and attempt to knock him off balance. Then she would hasten back to the side of her pup, and adopt a defensive attitude, baring her fangs and blowing with rage. Grand Jean struck her several times on the muzzle with his club but this only seemed to provoke her combative ardour.

Why he was so set on seizing this particular whitecoat when there were hundreds, thousands, perhaps tens of thousands of others in the vicinity, probably not even he could have said. But resolved to have it he was, and as he couldn't manage the feat alone, he called us to his aid. With three of us fending off the animal's charges, it wasn't long before we managed to drive her to the nearest lead, into which she slipped smoothly and easily, not so much out of fear as to escape the blows that were raining down on her poor spine.

But she would not leave the scene. She remained there, her head out of the water, watching us with her big vacant eyes, while Grand Jean killed, bled and sculped her magnificent pup. When the job was completed, she heaved herself out of the water and slid up to the spot where the bloody carcass lay on the ice. She tried to turn it over on its belly, prodding it with her muzzle. Then she lay down on her side, folding her flippers over her chest, in the hope perhaps that she could persuade the pup to feed. When she understood finally that there was nothing to be done, she raised her eyes and gazed sadly in our direction. There was no doubt in our minds that she was crying.

And I was reminded suddenly of the tears of the old man to whom I had given the liver the previous evening. The seal's sorrow had acquired a human dimension that was truly heart-rending, unbearable. What fools we were! The shrill whimpers of the whitecoats filling the air about us seemed to give voice to the sorrow of the inhabitants of the ice-floes, like the strident lamentations of crowds of mourners around the catafalque of a dead Sufi. Grand Jean looked at Camille and Camille looked at me and I looked at Grand Jean, and no words were needed to express our innermost feelings: the self-disgust, the shame, the scorn for our dastardly act – the vanity of human beings who think they have a monopoly on emotions and sentiments.

And still the animal wept. There was no doubt that she was suffering intensely. She was a truly exceptional creature, for most seals will make no move to defend their pups when attacked; the moment they sense the presence of a predator, whether man or polar bear, they abandon their young on the ice and throw themselves into the nearest water hole. Less than one per cent of the females will try to revive or suckle the corpse of a slain pup; and these are the very ones who have defended their young. The others do not seem to suffer unduly from the loss of their offspring; or if they do, they don't show it.

In any event, it was difficult for us to return to the business of clubbing seals after that lamentable incident: our hearts were no longer in it. Each time I raised my club to strike, I glanced quickly about me to be sure I was not being observed by some mother seal; and when I brought the club down, it was with the odd impression that the entire pagophile nation had its eyes on me.

This uneasiness, this sense of being an intruder whose every move and gesture was being observed, was accentuated by the fact that the sky was choked with clouds and the ice was a dull, leaden, ominous colour. The wind had abated and we were waiting for it to shift around to the other direction. The atmosphere was charged with an oppressive heaviness that weighed on us like a curse.

"We'll have to beat it when she starts to blow!" Camille said. "There's a real son of a bitch shaping up over there!" But his sombre predictions were not to be realized – not immediately at any rate. The proverbial calm that precedes the storm dragged on and on, as if to make us dread all the more what was to follow.

In the end, it was not the weather that brought us trouble.

When Nazaire and Arthur returned to fetch us, we loaded the dripping pelts into the skiff and were wasting no time in making our way back to land before the storm hit, sometimes even breaking into a run when the lay of the ice permitted. In our haste, we failed to notice a little pointed keg[4] jutting out from the vast stretch of flat ice we were crossing, and the *canot* hit it head on. We felt the jolt and quickly slackened our grip on the rope, but it was too late; the spike of ice had pierced the bottom of the boat and torn a hole several inches long in the thin plywood.

Some say that it is just these little vicissitudes that add a certain charm to the seal hunt; but there is nothing particularly amusing about them when they actually occur. After the inevitable stream of curses, oaths and imprecations, there comes a moment of deep despair, during which each man imagines the worst, though he takes care to keep his fears to himself. Then, to ease the tension, someone will begin to speak loudly about folks back on shore noticing that a team has failed to turn up and having the presence of mind to send a Government helicopter to its rescue before darkness. The men will then begin to feel a little better; one or two may even begin to joke about their predicament! But once the initial shock is over, they take careful stock of their situation and make plans for getting themselves out of their dilemma.

What it amounted to, in our case, was rendering the *flat* temporarily seaworthy. But this was easier said than done, for we had no tools but an old, Boy Scout axe that happened to be lying in the bottom of the boat and nothing whatever to use as a patch. Meanwhile, great gusts of snow were sweeping over the ice. Soon, the tide would begin to ebb and it would be impossible for us to regain land. The curse of the pagophiles was on us, I thought; it had been decreed that we should not be allowed to leave the kingdom of ice with the pelt of the abducted whitecoat!

When we informed Nazaire of our struggle with the mother seal, he was deeply upset. According to him, our behaviour was nothing less than sacrilege. We had broken the bond of fraternity between the seals and the men, symbolized by the blood we had drunk that morning from the heart of our first kill. "The pups the mothers defend are the princes of the seal nation," he told us solemnly. "You must never touch them. If you should kill one by accident or out of ignorance, then you must kill the mother, too. Without delay. Otherwise she'll suffer and communicate her sorrow to the entire herd."

Though I knew he was laying it on a bit thick, I thought there must be at least a grain of truth in his explanation.

Fortunately, destiny is not the slave of superstition and sorcery; there are also proverbs! The one about necessity being the mother of invention was to be confirmed on this occasion in a rather remarkable fashion, for Camille, a boat builder, suddenly had the brilliant idea of using one of the *canot*'s seats to patch the hole in the hull.

Very gingerly, he removed the seat, taking great care not to twist or lose any of the nails, which he practically had to draw out with his teeth for lack of proper tools. Then, he cut a large piece of material from a sweater, folded it in three, coated it with a mixture of motor oil and seal blubber, and used it to caulk the seams of his patch. He nailed the entire thing to the gaping wound in the hull, praying that it would hold.

And hold it did! Needless to say, the repair was not waterproof: a constant thread of water entered the boat, and none of us was very confident we would make it safely across the channel. But, bailing constantly and taking more and more pelts in tow, we reached the shore ice with the water lapping about our ankles. I truly believe that on that evening each one of us took a silent oath never to set foot on the ice-floes again.

But there are no drunkard's oaths less reliable than those of a sealer who has made first-hand contact with fear. How many times have I seen a team of sealers descend from a helicopter after spending a night on the ice, swearing on the heads of their women and children never to set foot on the floes again. The following year, at the first faint signs of spring, they are simply itching to be the first ones out.

The fear, you see, is followed by a sense of shame at being afraid, and this sentiment – in the case of men, at any rate – is generally stronger than the fear itself. You have to prove to yourself, and to the others who might suspect you of being afraid, that your courage has not deserted you. You reason thus: in the final analysis, one experiences fear before or after an event, but never during the actual circumstances that provoke it; otherwise, one would die of it. So fear possesses no fundamental reality. And the best way to convince yourself of that is to return to face it.

These ideas must surely have haunted Grand Jean, for the following day, while the storm raged and the winds howled, preventing all access to the ice, he set out to find us a new boat. A friend loaned him a dory that was in fair condition, and he spent the entire afternoon equipping it with a support to hold the motor.

Joining us that evening, he proudly announced that we were ready to return to the ice as soon as the weather cleared.

This was to be none too soon, however. We spent the next few days indulging in the pagan pleasures that are the lot of landlocked sailors. Then one day – it was a Sunday – returning from mass, Grand Jean's father was outraged to find us still lolling about the house, only half awake. It was a real shame, he said. The weather had cleared and all the able-bodied men of the Islands were out on the ice, killing seals right up to the beach along the West Dune. He'd seen them with his own eyes. And here we were, a gang of lazy good-for-nothings, lying about the house, wondering what to do with our big carcasses! Then his voice dropped a little: "You know that ice-floe you were working a few days back. Well, if I'm not mistaken, with the winds blowing like this, it must be out there right now in line with the church steeple." And, with his hand, he indicated a long stretch of water visible from the kitchen window.

A half hour later, we were headed in the direction of the floes glittering less than a mile out to sea. For three days, the storm had wreaked havoc with the ice, and all our reference points had vanished. Sitting at the helm, Grand Jean drove the dory through a wide lead and steered for the open sea, threading his way amongst the large ice pans that floated lazily in the current. Standing on the front seat like a foremast, Camille scanned the hummocks through his binoculars; aside from several birds that fled at our approach, flying flush with the water, all life seemed to have disappeared from the kingdom of ice.

It was only when we were a good half-dozen miles off shore that we began to hear the first faint echoes of the whitecoats. We circled several more empty floes, and then, suddenly, we were upon them.

But this was nothing like the patches we'd worked several days earlier. The entire area before us, from one end of the horizon to the other, was filled with ice-floes so thickly covered with pagophiles that they seemed about to sink beneath their weight. This was the main patch, the heart of the seal colony. The neighbouring waters seethed and bubbled where the black heads of hundreds of adult seals rose and fell, fizzing to the surface of the waves like bubbles of air in a gaseous liquid. No sooner had they blown and dived than they were replaced by others, and these by still others, an endless eruption of living periscopes rising and falling in the icy water about us. The sea was literally alive with seals.

On the ice, the whitecoats had altered considerably since our last visit: they were bigger now and many of them had begun to lose their white, silky, natal fleece. When we seized them by the scruff of the neck, large tufts of hair came away in our fingers. Some were farther along in moulting than others and already had black bald patches on their necks and about their flippers. Nazaire told us not to touch these seals, for their fur was no longer of any value. The hides of these *tanners*[5], he said, were good only for making leather.

There remained, however, a great number of pups who hadn't yet started to moult and whose fur was still firmly attached to the skin – enough to occupy our team and at least a dozen others, during that entire afternoon. But it was only a matter of days, if not hours, before these pups would also begin to moult, gradually discarding their bright foetal fur for a coat of thick, dark, speckled hair. The days of the seal hunt were nearing an end. From now on, many of the skins we would harvest would be of inferior quality.

Unlike most sealers, we planned to hold onto our pelts and to have them processed ourselves, selling only enough of the raw pelts to cover expenses. We all agreed that this was the only way to work the seals without being exploited by the formidable empire of the sealing industry and without unduly exploiting the herd itself.

What earthly sense was there, we reasoned, in a race that saw itself as the people of the seal stripping these animals of their fur only to let itself be stripped in turn by the boundless rapacity of big business? That day, while flaying the whitecoats, I dreamed of all the mittens, caps, vests, and moccasins I could make from their pelts, garments that are so soft and warm next to the skin on a cold winter's day.

Nazaire did not give the signal to return until late in the afternoon. When we set out for land, the tide had already been rising for several hours. We had harvested a great number of pelts and we were far from shore; there was no question of dragging that monstrously heavy dory over the ice. But the leads we had to thread to reach the main channel were already beginning to close up. Once again, the return to land filled us with apprehension.

At one point, we were making our way through a narrow lead between two huge floes that the tide was slowly pushing against each other. We had covered scarcely a quarter of a mile when suddenly the two floes came crashing together; less than a cable's length behind us, the formidable jaws of the ice-floe, studded with hummocks like the fangs of a giant dog, closed over the wake of our frail craft with a terrible explosion. It was at once a spectacular and a terrifying sight, for the point of impact of the two walls of ice seemed to be approaching us more quickly than we could flee it.

I suggested timidly that, while there was still time, we should lift the dory onto the ice and await the turning of the tide to regain land, rather than running the risk of being crushed in that colossal vice. But sitting at the helm, Grand Jean turned a deaf ear to this proposal. He had decided to carry on, and no plea would make him change his mind. He kept the throttle wide open, glancing first ahead, then behind, a serene look on his face, a trace of a smile on his lips.

But his smile vanished suddenly when he realized that the ice was moving in on us faster than we could escape it. The lead was closing up now no more than a few feet behind us. With what seemed an infinite gentleness, the two incredible masses of ice came together and locked; but the deafening sound of their impact and the accumulated debris that the shock impelled lava-like over the floes testified to the formidable force of the collision. We expected at any moment to be snatched up in those ominous jaws and silenced forever. There seemed no chance of escape, for even if we had leaped onto the ice-floe, we still would have been crushed almost instantaneously beneath the avalanche of crumbling ice. Meanwhile, the fatal moment was repeatedly postponed; each time it seemed to be upon us, the boat leaped forward as if by magic and escaped the impending doom.

Of course, there was nothing magic about this at all; the dory was only obeying natural laws. As they came together, the walls of the floes sent a large wave racing before them, similar to the swell caused by the prow of an ocean liner, and it was this that protected us from the monstrous pincers galloping in our wake. The moment the wave touched us, it picked us up and, with the aid of the boat's motor, propelled us forward at a great speed on the surf for a while. Then the boat seemed to founder for a moment or two before being seized again by the wave and carried once more out of danger.

I don't know if this terrifying pursuit lasted minutes or hours – it seemed like centuries – but I do know that all five of us were pale and trembling when we finally reached land. No one spoke, but we were all aware of our very close call; the slightest misfire of the motor, the smallest bubble of air in the fuel line, and the folks back home would have been talking about us in the past tense.

The following day, still very much under the cloud of our misadventure but trying not to let it show for fear of being taken for cowards, we were all very relieved to discover that a new storm had come up overnight. This time the winds were out of the west. For four days they blew without respite. When they finally abated and the Department of Fisheries' helicopter was able to make a reconnaissance flight, the herd had moved east of the archipelago and was drifting toward Cabot Strait, on its way out of the Gulf.

We returned again to the ice that spring, but less in the hope of finding seals than to overcome the fear that had set in following our narrow escape. The floes and leads were completely deserted. The adult seals, who normally remain in the area to breed and moult, had followed the caravan of their offspring out of the Gulf.

There was nothing left to do but to bury the pelts in the snow until it was time to sell or process them, and to store the *canot*, grapnels, clubs, ropes, and blood-spattered oilskins in the stable. The seal hunt was over for another year.

V
Days of Fever

It never lasts much longer than that: three or four days, a week at the most, just as long as the wind pushes the ice-pack up against our little string of islands and dunes. Nowadays, the seal herd rarely even comes within reach of the shallows surrounding the Islands; it floats by, far out to sea, oblivious of the prayers and entreaties of the Islanders. But each year the approach of the seals releases the same outbreak of fever amongst the local inhabitants, notwithstanding their realization that the herd may once again remain at large.

It's a strange thing, this fever, probably not fully comprehensible to inhabitants of more temperate or urban areas, for it is motivated by extremes of solitude and climate that do not lie within their experience. It is similar, I suspect, to the exhilaration the Eskimo feels when he first sees the dawn after the long polar night.

In the wintertime, on the Magdalen Islands, one also has the impression of being forsaken by the daylight. On those barren islets, prey to the winds and seas of the vast Gulf, the winter truly deserves to be called "the dead season."

The skies are choked with heavy, black clouds; but, even without them, the night would absorb two-thirds of the day. The winds howl incessantly down chimneys and whistle about the isolated houses scattered over the hillsides, shaking them to their foundations, splitting poles and pulling down electric power lines. The trawlers, back from the cod fishing grounds off Cape Breton, sit motionless at their moorings in the frozen harbours. All life seems to have come to a halt. The boats have stopped their runs to the mainland and planes take off only on rare clear days.

Isolated from the rest of the world, the Magdaleners endure these long months of darkness – December, which carries them right to the bottom of the year, and January, which painfully extricates them from it – by turning in on themselves and availing themselves of every opportunity for distraction. Anything but the burdensome inactivity that constantly weighs them down! The feast days of Christmas and New Year's usher in a season of rejoicing and festivity that lasts, as in pagan times, till the beginning of Lent.

In the old days, this was the marriage season. And what better moment could a young fisherman choose for his honeymoon than this period of long nights and unemployment? The man who took a bride during the fishing season never had such good fortune: his marriage vows were scarcely uttered when he was cruelly dragged from the nuptial bed by his skipper, who wanted him on the deck of the trawler and not in the arms of his wife. But winter was the ideal time for a wedding; then, even the skipper would join in the festivities.

The celebrations often lasted an entire week. Several days before the ceremony, while the prospective groom was busy saying goodbye to his bachelor life, the relatives would begin to arrive in bell-jingling sleighs. They came from all the islands, and, since such visits were rare, they made the most of them. In those days, there were neither roads nor bridges, and travel from island to island could be hazarded only when the seas were frozen. In all likelihood, they would not see one another again until the following winter.

Visitors were lodged without fuss in neighbouring houses, and while the women bustled about making last-minute preparations for the marriage feast, a flood of children running in and out among their legs, the men sat huddled around a small pot-belly stove in the shed, talking gravely of the sea, their pipes stuck in their mouths, their long gnarled hands spread out on their thighs. In the evenings, they played cards by the light of an oil lamp; then the candles would be snuffed out while someone told tales of the sea.

But it was after the wedding celebration, when the priest had taken leave of the guests, that the real festivities began. The fiddles were taken out of their cases, the jugs of *bagosse*[6] and the flasks of *chien*[7] were dug out of the dark corners where they'd been hidden, and the furniture was pushed back to make room for the dancers.

There were reels and rigadoons, square dances and farandoles. The fancy steppers competed with each other, jigging to the music like double-jointed marionettes, executing steps that were more and more complicated, more and more rapid, their eyes fixed intently upon the fiddlers, their faces streaming with sweat; while the old folks, sitting bolt upright on their chairs, tapped their feet or played the spoons on their knees. Then, it was the turn of the singers and storytellers, who kept their audience captivated until the wee hours of the morning.

The next day, the festivities would move to another house in the village, and, the day after, to yet another; and so it would go, until either weariness overtook the revellers or the troubadours had to leave to entertain at another wedding. In the old days, during the winter season, the fiddlers never stopped fiddling.

Today, things haven't changed much. Of course, traditional weddings have now been supplanted by stereotyped parish carnivals, and the celebrating has moved away from the home into bars, community halls, and other public places where electric guitars tend to bury beneath their heavy, reverberent chords the melody of the fiddles. But the folkloric nature of the festivities remains rich and alive, and the celebration still serves as an antidote to the boredom and emptiness of the dead season.

By Candlemas, February 2, the Magdaleners begin to hanker for a change of pace. After all, celebrations are fine as long as there is something to celebrate, but they begin to lose their appeal when the days become a merciless repetition of unrestrained revelry. The women begin to nag their husbands to let up on the drinking – "A fine example for the children!" – and the men become irritated with themselves for being so lax.

The anticipated arrival of the seals acts as a remedy. At last, there is a sign of activity on the horizon, something for the people to look forward to in this disorganized life of theirs. The men begin to look upon their clubs, grapnels, and ropes as more than mere decorations hanging from a nail in the barn. They draw the knife from its wooden sheath and test the blade on the fat of the thumb. The seals become almost tangible, almost within reach. Soon, from all the stables on the island, there is heard the shrill, rhythmical songs of knives being sharpened on the steel. The indefatigable wheel of the seasons has once again begun to turn.

The men of l'Anse-à-la-Cabane and l'Etang-des-Caps have already begun to harvest the *rouards* at Deadman Island. These are the pups of the grey or horse-head seals, a large colony of which inhabits this rocky islet a dozen miles southwest of the archipelago. Born in mid-January, they resemble whitecoats except that their foetal fur is sparsely dotted with reddish-brown spots. This hunt acts as a powerful detonator for the explosion of fever prompted by the anticipation of the arrival of the great pagophile herds. The seal is no longer only a dream, a disembodied hope; it has suddenly become a reality. The sealers make the rounds of the parish, selling flippers, carcasses and livers, and the first meals of flipper stew whet appetites for the bounty of fresh seal meat to come.

At all the wakes, in all the corner stores, all they talk about is the seals. The old men draw upon their memories of the seal hunt in days long past: "Did I kill whitecoats?" they exclaim. "You better believe I did, my boy! I killed more than you'll ever clap eyes on! I wasn't no slacker, let me tell you!"

Even taking into consideration the inevitable amount of exaggeration, which merely adds charm to their glorious tales, one is left in awe at the mere thought of the extraordinary number of seals that inundated the Gulf in those days. Each spring they came without fail, swarming in waves along the dune, from West Cape to Cap du Dauphin; and with a strong nor'wester blowing, they'd even come right up to the front doorsteps.

Daniel à Claude, ninety-two years old, recalls his first hunt: "In those days, we took things pretty much as they came. I was fifteen at the time, I recall, and I'd gone out on the ice with my brother. We had no *canot*, no grapnels, no clubs – just a knife and a coil of rope. The seals were swarming all over us. I'd grab one and *wham*! punch him hard in the snout. I killed 'em with my fist. Then I carried 'em over to my brother who sculped 'em. Seventy-five skins we brought back that day! When I got home, my hand was swollen big as an elephant's foot, all red and purple it was, and it hurt like hell. But the next day, I went out again and kept on punching 'em out!"

Each year, the young people listen with awe to the same stories, and if the eloquence of the old men begins to wane, it can always be rekindled with the same question: "How come we don't see so many seals today?" And the old men gravely shake their heads and reply: "It's because of the massacres on the east shores of

Newfoundland. That's what it is. We were too few ever to do them much harm."

"Except during the time of the airplanes," someone of the middle generation will then be heard to say, but quietly, gently, as if to make it clear that he has no intention of contradicting the old man's words. For the wisdom of age is still held in high regard in the Magdalen Islands.

Then the old man will continue, in the rich idiom of the region: "Oh, it weren't human in those days! Every spring, the airplanes arrived, thirty or more of 'em, Pipers, Cessnas, you name it – some in such a sorry state they could hardly get off the ground. But the pilots were men, I'll grant you that. They set their machines down on ice-floes no bigger'n from here to Adrien's place across the way. They didn't give a damn what was under 'em, slob ice, hummocks, snow, when they made up their minds to set down, they set down. When they were ready to land, they'd turn to us and say: 'Hold tight, boys, here we go!' There was nothing to fear, they said. If we didn't make it, we sure as hell wouldn't be around to tell about it!

"They were real daredevils, those pilots! And we were fool enough to go up with them! We'd be all crammed together, four or five in a Cessna, three in a Piper, and away we went! Oh, there were accidents, sure. But we brought back a heap of sculps those years. Thirty, forty thousand every spring, by plane alone. Then there were the big boats from Norway and Nova Scotia that worked the herd, too. Yeah, I reckon you could call that a sort of massacre, too. Right here around the Islands."

In another township, in another kitchen, seated around another table, other men will also be talking about the seals. It's as if, for a time, no other topic merits consideration. From being a sort of therapy, the seals have become an obsession. Here they recount the tragedies that befall the men on the ice: blizzards that come up out of nowhere, winds that turn, bodies that freeze. There they heatedly discuss the prices the pelts will fetch this year, for already the buyers have circulated staggering rumours to incite even the most reluctant sealer to join in the hunt. For this, too, is part of the folklore of the hunt: not until the pelts are ashore will the prices begin to drop.

Somewhere else, without fail, in a dark corner at the back of some tavern, a couple of tipsy sealers will be gulling some credulous listener with their debate about the enigma of the whitecoat's sex – a controversy as old as the seal hunt itself. How is it, they wonder aloud, that all male whitecoats are born in the waters off the Magdalen Islands, and all female whitecoats on the eastern shores of Newfoundland?

The listener is dumfounded: this goes against all the laws of probability. Nevertheless, it's true, they maintain; they're ready to take an oath on it: each time they cut open a whitecoat – well, at least ninety-eight per cent of the time – the knife strikes a small pointed bone in the region of the lower belly. What can that be but the penis? Then they have to stop and regrind their knives before continuing with the sculping. What better proof that there are only male whitecoats in the Gulf herd? And doesn't it stand to reason, then, that all the females must be born on the east coast of Newfoundland?

At first sceptical, the listener is finally obliged to concede the point, for the men's observations can be substantiated by any number of sealers. But even more intriguing is the fact that the Newfoundland sealers are also convinced, for their part, that all male whitecoats are born on their shores and all female whitecoats in the Gulf. And why? Precisely because of this same little bone upon which they, too, blunt their knives ninety-eight per cent of the time!

Does this mean, then, that there are no female seals? Are the animals hermaphroditic?

I must admit that I, too, taxed my brains for some years over this dilemma. Whenever I had the opportunity to speak to a biologist, I would ask him to clarify the matter, but the most I ever got for my efforts was an ironic smile and the unequivocal reply that there was no truth whatever to the sealers' contention: there were as many females as males in each herd. To tell the truth, I was more than a little annoyed with them for their condescending manner, the way they sneered from their lofty seats of learning at the unanimous observations of thousands of sealers, corroborated by my own personal experiences on the ice. I was on the point of raising the question before the Academy of Sciences when I came upon a text of natural history which, in a few lines and a couple of sketches, cleared up the entire mystery.

The whole matter is one of male chauvinism, pure and simple. Like so many men who consider the female sex organs virtually non-existent, the sealer naturally cannot imagine that the little pointed bone is anything but an attribute of male virility. However, nature has seen to it that, in the case of the pagophiles, equality of the sexes is not merely a catch phrase: if the males are endowed with an unusual penis bone, the females in turn possess a clitoris bone that is every bit as remarkable!

Thus, throughout the Islands, thousands of conversations, some serious, some frivolous, centre on the seals, building to the pitch of an incantation. Meanwhile, in the boat sheds, the *canots* are being constructed. The ribs of steamed oak are pressed over the wooden frames of the hulls, and caressing these skeletons, the sealers dream aloud of their performance on the ice: will they manoeuvre easily over the hummocks? Will they slide smoothly over the ice? . . . Already, they can see them filled to the gunwales with pelts! They live on their hopes, and the fever mounts steadily with the waxing of the last winter moon.

And it is generally a kindly moon. Cool sunny days succeed the violent storms of early winter, and the landscape of the archipelago offers an incomparable spectacle of naked buttes glazed by the whistling wind: sensuous Maillol breasts sparkling in the blinding daylight or throwing bronze flames on clear, moonlit nights. The Magdaleners are passionately enamoured of their environment, and there is no doubt that its rediscovered beauty, during the days of this February moon, acts like a tonic, colouring their entire attitude to life. The air has a crystalline, depthless quality to it, the light makes the iris of the eyes dilate, and the already strong sun – similar to the Saint Martin's summer sun – causes a new ardour to rise in the veins of the local inhabitants like sap.

Seated beside their stoves, the women watch over the molasses biscuits they are preparing for their men to take out on the ice, that mysterious world they know so little about, but fear with all their hearts because of the terrible tragedies that have befallen the men out there. They worry themselves sick, anticipating the worst, praying that the herd will drift far out to sea, far beyond the point where even the hardiest sealer will venture in pursuit. Some recite rosaries. "Keep him safe!" But privately they know they will never be able to bring themselves to ask their men to give up the hunt, so deep is the tradition in their bones.

The young lads have no such fears! The tales of their elders fire their imaginations. For a time, the hockey stars are eclipsed by the heroes of the hour: Wellie à Daniel, Jimmy from l'Echouerie, Nathaël à Evariste – the great seal hunters. The boys harness themselves to *canots* and race about the stables, pulling the boats behind them. And when they can escape the eye of their elders, they rush down to the capes and play at copying, jumping from ice pan to ice pan, just as they've heard tell real hunters do.

I've never seen them do this, for they take great care to keep well out of sight while at this forbidden sport, but invariably each year one or another of these little rascals falls into the water and gets soaked to the skin. But not even that deters them; these children of nature are immune to just about anything! They merely imitate their elders, wring out their wet clothing, slip back into it, and continue with their games, hoping they'll be dry before it's time to go home. For what they fear, above all, is the punishment awaiting them if they return to the house with wet clothing.

One learns early the art of seal hunting in the Islands. My daughter Dominique was only four years old when she began to understand what it was all about. For several weeks, she had been listening to us talk incessantly about the ice, without once being allowed to put in a single word. Then, one day, while her Uncle Grand Jean and I were discussing the inevitable subject over a jug of *bagosse*, she came up to us, dragging my club in one hand and holding her little plush seal in the other. Announcing in a loud, ringing voice that she was going to show us how it was done, she set the seal on the floor, raised the heavy wooden club in both hands and *wham*! let it fall with a thud on the head of her toy. "You see, I can do it, too!" she declared triumphantly. And she proudly informed us that she was going to go with us the next time we went out to the ice.

At the approach of the full moon, the fever reaches its peak. Somewhere between the Magdalen Islands and the Gaspé Peninsula, the female seals have heaved themselves onto the hard ice being disgorged by the river to whelp, and by an ineffable osmosis or telepathy, the more experienced sealers get wind of it. One fine morning, sniffing the air at the water's edge, they discover a slightly different odour to the atmosphere, an indefinable, almost imperceptible scent that wasn't there the day before. "So," they say, "the females have taken to the ice." And they are never wrong.

In the old days – in *l'en-premier*, as the Acadians say – this was the time when the hunting parties set out for the ice. In teams of three, with a small *canot* as light as a feather for crossing the waterways, but without clubs or knives, they would travel fifteen, sometimes twenty miles from shore – for no reason other than to look the area over. The first few times out, it wasn't unusual for them to see nothing but virgin ice-floes twinkling all the way to the horizon. Then they would retrace their steps, satisfied with their "promenade," as they modestly called these long and perilous excursions, and wait to set out again another day.

When they spotted the seals, they hurried back to the village to awaken the beadle, who would ring all the bells. A volley would be fired at the end of the sand dunes to advise the other parishes that the herd was on their doorstep, and soon, from various points in the distance, all the bells of the archipelago could be heard pealing out the joyful news.

Today, the helicopter has supplanted the long excursions on foot, and the telephone has replaced the old ships' bells that were used in our earliest churches. But the passage of time hasn't diminished the euphoria that grips the population once the seals are finally sighted. There is electricity in the air, a great wave of hope and delirium that washes over the Islands. It's as if the Magdaleners had once again been assured of their survival!

To appreciate this excitement fully, you would have to experience it, at the end of winter, after that long period of solitude in the middle of the vast Gulf. You would have to have known the monotony of eating salt fish and corned meat day after day to know how the mouth waters at the mere thought of fresh seal meat. You would have to have seen the beaming faces of the children when they're served their first meal of fried whitecoat liver, to have participated in those gluttonous banquets around a cauldron of flipper stew, to comprehend how deep are the bonds that, for generations, have united the little island people and the seals . . .

For it must not be forgotten that it was largely because of the harp seal, that miraculous little creature, that several hundred Acadian refugees, hunted down, harrassed, hounded over the globe, were able finally to settle and survive on these windy islands, despite the misery and slavery they were subject to during the troubled second half of the eighteenth century when the riff-raff of the British armies celebrated, with murder, carnage, and genocide, their victory over the Indian and French-Canadian nations, who had preceded them on the shores of the Gulf and the St. Lawrence River.

We moved in the direction taken by the crows, hauling the light skiff over hummocks, ice fields, and rock-hard billows of snow.

The beauty of this sudden intense abundance of life took us by surprise. We stood, feasting our eyes on the amazing tableau.

The plaintive sound of the whitecoats is like a lament woven with tears, wails, and whines that merge into an incredible chorale. It is said that a mother recognizes her pup by its call.

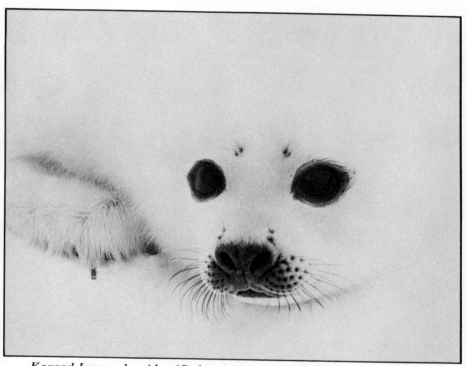

Konrad Lorenz has identified certain physiological characteristics that trigger feelings of protectiveness. The shape of a "baby" seal, like that of a child, makes us automatically feel sympathy for it, whether or not it wants or needs our attention and care.

The body stiffened as if to protest against the blow. It was shaken by a few convulsions, and then lay still. It was dead.

Removing the skin of a seal is a delicate and skilful operation. You can't just tear it off as some self-styled witnesses claim.

We retraced our steps over miles of hummocks and pressure ridges, straining to drag half a ton of fur, fat, and meat over the ice.

We men are simple beasts, unhypocritically accepting the harsh law of nature: one must kill in order to survive.

This year, as in years past, the Magdaleners will come out to welcome the men home from the hunt.

A week, ten days at most, and the hunt is over for another year.

VI
The Odyssey of the Magdaleners

I have never liked beginning a story at the beginning: first, because it seems rash to affirm that it begins at one particular point; and, second, because such an approach obliges it to move at a pace that is constipated by chronology, preventing it from opening out in all directions, and trussing it up like a madman in a strait-jacket. I might just as easily have begun my account on the islands of Saint-Pierre and Miquelon, about mid-April, 1793.

Abbé Jean-Baptiste Allain was waiting for an east wind. But again, that morning, there were flurries out of the northwest that lashed his cheeks the moment he set foot out of the presbytery to go and say early mass. He felt a vague despair, which he immediately suppressed.

It was not for lack of devotions, prayers, masses, or even rigorous fasts verging on holy mortification, however, that the All Powerful had refused to grant his wish. He had made more than his share of sacrifices. And time was running out. He thanked the Lord once again for the test He had put him to by allowing the wind to blow in the wrong direction, and then once more fervently implored Him for an easterly breeze.

Since the previous autumn, the good priest had been torn with anxiety. The news brought by the last ships from the mother country – an expression that he thoroughly deplored – had not been at all reassuring. For his refusal to ratify the Civil Constitution on the Clergy, Louis XVI had been imprisoned, then deposed, and the Legislative Assembly, now known as the Convention, had subsequently proclaimed Year I of the Republic.

Communications being what they were, there was no way for

Allain to know that the head of citizen Capet[8] had dropped into a burlap sack to the exultant cheers of the Parisian populace on January 21, 1793, nor that his execution had inaugurated a season of intense activity on the scaffolds of the bloody guillotine, which would later aptly be called the Reign of Terror. Scores of individuals from all levels of society would pass that way, to the sound of the carmagnole: the king, his Austrian wife, nobles, bishops, priests, monks, and ordinary citizens, first the uncommitted and the moderates, then the more ardent and fanatical supporters of the Revolution; until finally, on Thermidor 9th of Year II, the mastermind of all this madness, Robespierre himself, would take his turn at placing his neck beneath the sharp blade of his own invention.

Though Allain knew nothing of these dire events, he faced the future with great misgiving. He dreaded, above all, the arrival of the first boat from France. Would it not be filled with soldiers and civil servants of the new regime, and would it not be necessary for the people of Saint-Pierre and Miquelon to swear allegiance to the ignoble Republic? Picturing priests slaughtered in their cells, he decided he must flee before it was too late.

For three years now, the curé of Miquelon had been a fugitive. A native of Granville, on the Gulf of Saint-Malo, born into a family whose influence had extended beyond the borders of Maine and Calvados, right into the velvet-hung antechambers of the Palace of Versailles, he had prudently sailed for England at the first faint rumblings of the Revolution, in the company of Jean Barthélémy Hubert, the nephew of Mgr. Jean-François Hubert, Bishop of Quebec.

From England, through the intervention of Mgr. Hubert's curate, who was then serving as chaplain at the French Embassy in London, the two men managed to steer a course for Saint-Pierre and Miquelon, the last remaining French stronghold in North America. And from there, protected by their isolation, they had followed with growing concern the political developments in Paris.

As the situation deteriorated, they devised the audacious project of emigrating to the Magdalen Islands and tried to induce the inhabitants of Miquelon, all of Acadian stock, to follow them in their wild adventure. But, weary of being shunted from place to place for nearly forty years, and finding themselves momentarily at peace in the green hills of Miquelon, the Acadians were not enthusiastic about the plan. The sea has no secrets from those who make their living off it, and they knew that the Magdalen Islands belonged to Richard Gridley, a Boston tyrant who had fought with General Wolfe on the Plains of Abraham. The mere thought of kneeling before this vile creature to take the infamous oath of allegiance to His Majesty, the King of England, filled them with disgust.

Wasn't it their obstinate refusal to take this oath of allegiance and to carry arms against the French and their long-time allies, the Mic-Macs, that had caused them to be hounded out of Acadia in 1755? The wounds in their hearts were still too fresh not to inspire them with hatred.

Voices broken with emotion, the old people who had lived through the Expulsion, *le Grand Dérangement*, recalled the terrible events that had followed the reading of the Order of Deportation in the church of Port-Royal, on September 5, 1755. The previous night, the troops of Edward Cornwallis, the Governor of Halifax, had begun to invade the farms, setting fire to the crops, slaughtering the cattle, sowing panic in the stables, killing, mutilating, and torturing anyone who offered the slightest resistance, with no regard whatever for sex or age.

Those who failed to escape to the woods were taken prisoner and penned like beasts in enclosures guarded by men with muskets, while English infantrymen pillaged the prosperous farms the Acadians had carved out of the swamps of the Annapolis Valley during the previous century. Their only crime had been the desire to remain neutral.

Soon, the roads were clogged with long lines of deportees, whom the British soldiers drove on with sticks and cudgels, deaf to their pleas and entreaties, herding them toward the loading docks and the boats that waited to carry them far from their native shores. In the crush, mothers were separated from their children and husbands from their wives; but far from allowing them to seek out the lost ones, their captors seemed merely to take a malicious pleasure in further increasing their misery and confusion with fresh insults and blows.

Abbé Allain must have heard this story a hundred times! And he knew that it was not only for military or political reasons that this nation of twenty thousand souls had been routed from its homeland. Though they would never admit it, the English had coveted those prosperous farms.

Since their arrival in America at the beginning of the seventeenth century, the valiant emigrants from Berri and Poitou who were later to be known as the Acadians, had transformed the vast swamp of the Annapolis Valley, flooded twice a day by the twenty-metre tides of the Bay of Fundy, into a rich, fertile, alluvial plain. This had been accomplished with the aid of *aboiteaux*, a series of ingenious dykes that held back the waters of the tidal bore during high tide and allowed the land to drain during low tide. The settlers had planted orchards and vegetable gardens and had grown wheat and other grain foods; at the time of the Expulsion they had possessed more livestock than Nova Scotia – as their lost Acadia was later renamed – would ever again be able to boast.

Where the Acadians differed most from the other colonists who tilled the soil of North America, and particularly from the racist Anglo-Saxons, was in their friendly relations with the native peoples. Scrupulously respectful of the Indians' rights, traditions, and customs, they were not content merely to settle on stretches of land that were previously uninhabited; they also cultivated bonds of cordiality and friendship with the natives that even resulted, on occasion, in inter-racial marriages, to the great consternation of the apostolic Roman Catholic Church.

Despite their deep and abiding faith, the Acadians left the work of evangelization to the missionaries and abstained from all proselytizing. On the other hand, they proved extremely receptive to the naturalist philosophy of the Mic-Macs. They did not hesitate to participate in native rites and ceremonies, and they considered the Indians to be their equal in all respects. As they felled trees to build their houses and to provide them with heat – a practice which the Indians did not share and at which they must surely have looked askance – they got into the habit of replanting as they went. They killed no more game than they could consume; they venerated the earth and the forest with a devotion that would have been the delight of present-day ecologists; and there is no doubt that their small, peaceful, hard-working community, steeped in the widsom of the "savages," was as close to the traditional concept of Eden as anything this continent has ever seen.

Cornwallis' soldiers herded the Acadians into the dark, damp holds of the ships, with no concern whatever for family ties. Grief-stricken mothers watched their children – even unweaned babes cruelly torn from their arms – vanish into the black holds, manhandled by the soldiers. In most cases this was the last they would ever see of them. In vain did they struggle, moan, and protest; their cries were drowned in a din of wailing children, clanking rifles, and baying orders. The hatches were shut on this human cargo, the sails were raised, and the ships set sail with a creaking of winches and ropes.

Then it was their turn to embark upon other vessels and to experience the gloomy, asphyxiating atmosphere of the holds, with hunger, thirst, cold, and harrassment as their constant companions during the long weeks of the voyage, lying in their own vomit and excrement while rats scurried around their feet, succumbing one by one to cholera.

Where were the deportees bound? Great care was taken not to inform them, but England had arranged for the Acadians to be scattered to the four winds so that they would never again find their way back to their native shores. The majority of the exiles were set down on the coasts of Maine, the Carolinas, and Virginia, where the ships put into port. Others were carried as far as the United Kingdom, Holland, and France; still others to Guiana and Labrador. Most of the half-breeds were sold as slaves on the auction blocks of Charleston and Jacksonville, for many of the captains whose boats had been requisitioned for the deportation were slave traders.

Never would Abbé Allain and Jean Barthélémy Hubert succeed in convincing their parishioners that the ungodly regime that had recently come to power in France could be more barbarous than that.

Those Acadians who would later make their way to the islands of Saint-Pierre and Miquelon and who would still later become the Magdaleners, were set down on the shores of the thirteen American colonies, without food or supplies, with only the filthy rags they had on their backs. They wandered from village to village, spurned like the plague, in search of the scattered fragments of their large families. At the crossroads, they occasionally encountered other little groups of exiles who, like themselves, were tramping the roads in the hope of finding children, parents, friends lost since the Expulsion. They exchanged bits of news, wept over their lost homeland, shared scraps of food, and then went on their way, weary and disconsolate.

Some families, of course, were reunited; but, depending on the number of years that had passed, these *retrouvailles* often resulted as much in disappointment as in relief. A man, believing his wife dead, had re-married; children, separated from their parents, had been adopted by another family; teen-agers, last seen as young girls, had become mothers. Tears of joy mingled with sobs of bitterness and regret. There were outbursts, confrontations; but these brave people had suffered enough to know the real meaning of mercy. They had become fatalistic, and they forgave all, leaving to misery the discord that misery begets.

Thus, in the early 1760s, several hundred Acadians – men, women, and children, young and old – came together on a Carolina beach bordered with tamarack and scarlet oak. Their names were Sire, Hébert, Petit-Pas, Le Blanc, Gallant, Miousse, and they were diehard patriots, every one of them.

Returning to Acadia and swearing the oath of allegiance in the hope of recovering their property, as some of their compatriots were prepared to do, was as far from their minds as settling down in Canada, Louisiana, Maine or New Hampshire. They flaunted their resistance! For them, the fall of their homeland was no more than one lost battle in the all-out war between French and British forces for the North-American continent. They would regain the fortress of Saint-Pierre and Miquelon, and from there reconquer their land.

With their makeshift tools, they set to work like Titans, building the boats that would take them home. And, by the end of the autumn of 1763, they were ready to depart.

The season was not a propitious one for a long voyage, and the Acadians were not yet the accomplished seafarers they were later to become. Their knowledge of navigation was limited to a few rudimentary principles. Nevertheless, the prospect of a long voyage over the stormy seas of the North Atlantic did not fill them with undue alarm; they had enough experience of misfortune to know that the sea could do them no more harm than they had already suffered at the hands of their fellow men. So they set out confidently, steering by the north star, braving the waves and storms, relying on fish for their daily sustenance – the castaways of history and destiny riding out the deluge in their frail arks.

They reached Saint-Pierre on Christmas morning, and legend has it that their first act upon setting foot on land was to throw themselves at the feet of the priest, Abbé Lejamtel, beseeching him to bless the not quite legitimate marriages that had taken place during their long wanderings, and to baptize the children that had resulted from these unsanctified unions. For eight years, these wretched creatures had survived without the comfort of religion, and in constant fear of divine retribution.

The official registers of Saint-Pierre for this period are filled with entries like the following: "Hébert, Anastasie, daughter of Magloire and Anne Cyr, born April 6, 1756, Charleston, South Carolina, baptized December 26, 1763, Saint-Pierre; formerly of Acadia, deported to New England . . . Cormier, Séverin, son of Vilbon and Marie-Céleste Petit-Pas, born November 29, 1759, Quincy, Massachusetts . . ." By 1765, 105 Acadian families – more than five hundred people – had made their way back to Saint-Pierre and Miquelon.

Life went on. France, increasingly pushed about on the North-American chess board, refused to sanction the audacious project of reconquering the homeland, and the Acadians were resigned to respecting the armistice. They built little farmhouses in the verdant valleys of Miquelon, and spent their days farming, raising cattle, and fishing. Their lives were peaceful, punctuated by the faint ringing of the angelus calling the faithful to devotions. The pastoral quietude soothed the wounds of the Expulsion like a balm. The people even learned to be happy again in the simple round of their daily lives.

But this tranquility was no more than a mirage, for the future still reserved more than its share of misery for these hapless refugees.

One morning, in the autumn of 1776, the population of Saint-Pierre was rudely awakened by the sounds of a cannonade, followed by the loud ringing of bells. At the entrance to the harbour, an English frigate was bombarding the town with shot from its forty-eight guns. The fortress's battery, wiped out by the first volley, lay silent. Taking refuge on the heights, beyond the range of the cannon balls, the citizens gazed in horror as the buildings of their little town collapsed and, one by one, burst into flames.

Leaning on the rails of his vessel, Rear-Admiral Montaigue savoured his vengeance against his most hated enemy, France. Vexed by the recent bitter and decisive defeat of the British fleet at the hands of the American revolutionaries, he was resolved to make France pay dearly for the aid that the regiments of La

Fayette and Rochambeau had brought to the troops of George Washington, and he found his opportunity for an easy victory on the route of his inglorious return. Intoxicated by the odour of gunpowder and the power of the explosives, he calmly watched the town go up in flames.

He didn't silence his artillery until there remained nothing of Saint-Pierre but a heap of smoking rubble. But his thirst for vengeance remained unquenched. In the days that followed, he deployed bands of Newfoundland plunderers from Fortune Bay to sack and burn the little farms of Miquelon. And the two thousand colonists of the archipelago, one-third of them Acadians, found themselves at the outset of winter without homes, provisions or livestock, utterly destitute.

France had no choice but to repatriate them.

Once again, the Acadians were packed into the dark holds of brigs and frigates for the long and agonizing voyage over the rough waters of the North Atlantic.

The citizens of Granville did not greet the Acadians with great enthusiasm. The harvest had been a poor one in Britanny that year, and the prospect of having to share their meagre resources with two thousand famished refugees did not at all please them. Fortunately, the Clergy, moved by Abbé Lejamtel's accounts of the extraordinary piety of the refugees, came to their aid. But as time went by, and the Acadians, expecting any day to be returned to America, refused to assimilate with the local populace, even the Clergy began to find their presence something of a burden.

Seven long years had passed when King Charles of Spain requested his counterpart, Louis of France, to turn his Acadian colonists over to him so he could send them to Louisiana, which had fallen temporarily into his lap and which was the homeland of a number of other Acadian deportees. But most of the repatriated inhabitants of Saint-Pierre and Miquelon wanted no part of this plan. This was not the corner of America to which they wished to return. Would the French never understand? They left it to those Acadians who had been deported straight to France at the time of the Expulsion to offer themselves as volunteers for Louisiana.

It was during the exile in Britanny that Abbé Jean-Baptiste Allain first came in contact with the Acadians, whom, many years later, he would lead to the Magdalen Islands. This young priest was in charge of the Church's charitable foundations in Granville, and his duties put him in constant contact with these disinherited children of history. He had been deeply moved by the tales of their dispersion, their misfortunes, and their sufferings; and he felt a boundless admiration for the faith, courage, and humility that still animated them after so many reversals and mishaps. In his mind, these men, women, and children were kin, through their martyrdom, to Christ himself. Inflamed by their holy zeal, he fervently prayed that the All Powerful might make it possible for them to return to their beloved Acadia.

On a more practical level, he used his influence to gain audiences with people in high places, with a view to hastening their return to Miquelon.

But there was to be no voyage for Abbé Allain in 1783. Although Louis XVI had recently promulgated the Act of Recolonization of Saint-Pierre and Miquelon, following the signing of the Treaty of Versailles, which guaranteed the independence of the United States and restored peace to the North-American continent, the French economy was in such a wretched state that France could afford to send no more than 432 colonists – about half of them Acadians – to America. When Abbé Lejamtel was selected to accompany them to tend to the needs of their souls, Abbé Allain found himself on the quay of Saint-Malo on the day of their departure, standing among a crowd of idlers, parents, friends, Bretons, and Acadians, who had come to see them off and to bid farewell to the dear ones they would probably never see again.

Armed with their indomitable courage and their ingenuity, the Acadians set to work to reconstruct the town of Saint-Pierre and to rebuild their little farms in the valleys of Miquelon. But no sooner had they completed this formidable task than new events began to take shape on the horizon, events that would once again play havoc with their existence. In Paris, the mob had stormed the Bastille and had invaded the Tuilleries, crowning the king with the Phrygian cap.

When news of these events reached the distant colony, there was a polarization of opinion: the French colonists applauded these daring blows struck against the absolute power of the monarchy, but the Acadians were appalled. For them, democracy was essentially a British system, an invention of the Walpoles and the Pitts, and they had had a taste of the abuse of that institution at the time of the Expulsion. Their distrust, however, was directed only against the term *democracy*, for long before Voltaire and Rousseau had popularized the principles of equality, liberty, and fraternity, the Acadians were practising these rules in their everyday lives.

Both diplomat and pragmatist, Abbé Lejamtel skilfully manoeuvred to maintain peace and cordiality amongst the members of his flock, observing a prudent neutrality and endeavouring to view the Revolution that had broken out in Paris with a touch of philosophical irony: "You know the French," he told the Acadians. "They're short-tempered, chauvinistic, and extremist, but their tempers never flare up for long." And to the Bretons, he said: "You'll see, Mirabeau and Danton will persuade the king to take a more conciliatory stand toward the Third Estate, and Monsieur de Necker will manage to put the royal finances back in order. Then things will return to normal."

It took only the arrival of Abbé Allain and his companion, Jean Barthélémy Hubert, late in 1791, to upset the delicate peace.

At Miquelon, the two refugees had regained contact with the Acadian community, and they had wasted no time in outlining details of their projected flight. But not even their alarming reports of events transpiring in France succeeded in shaking the loyalty of the Acadians to the mother country; no more than did their grandiloquent speeches on the marvelous prospects for peace and prosperity in the Magdalen Islands. Until they were better informed, the Acadians had a profound distrust of the conditions that would be their lot in a life under the iron rod of a tyrant like Richard Gridley.

So, deciding on another approach the two men began to speak of the bright prospects of relocating members of their large dispersed families in the Magdalen Islands and of rebuilding their lost Acadia there together.

Thaddée Snault, nicknamed le Grand Snault because of his Herculean stature, was one of the first to nibble at this bait. A native of Ile Saint-Jean (now Prince Edward Island), he had been only ten years old at the time of the Expulsion, and he had spent his entire youth wandering about the colonies of New England in search of his family. He had finally found himself on that Carolina beach bordered with tamarack and scarlet oak where the Acadians had built the boats to take them to Saint-Pierre and Miquelon. There, he had fallen in love with a young orphan by the name of Evangeline Boudreau, whom he had subsequently married, and who had given him six fine children. Yet, despite his newfound happiness, he still nourished a feeling of nostalgia for his lost homeland – the deep green forests, the red earth and the golden sandy beaches of his native isle – and he had never lost hope of one day being reunited with his family, even after forty years of anguished separation.

Taciturn, wise, a man of great physical and moral strength, le Grand Snault exercised a considerable influence over the other Acadians of Miquelon. Abbé Allain, therefore, assiduously cultivated the friendship he enjoyed with this remarkable man. Often, by the light of an oil lamp in Snault's little house that shook and trembled in the winds, they talked late into the night.

In the spring of 1792, Barthélémy Hubert left for Quebec to meet with his uncle, the Bishop, and to obtain precise information about the Magdalen Islands. When he returned in the autumn, he was in possession of the arguments that would seal the fate of the little Miquelon community.

The archipelago in the Gulf, he reported, was inhabited by some twenty Acadian families and five Canadian families, all of whom had taken the oath of allegiance to His Majesty, King George III of England, on August 30, 1763. A French missionary, Abbé Leroux, had visited them several times since 1774. They made their living hunting walrus for Colonel Gridley. A number of Indians were still engaged as their assistants at the time of the missionary's last visit.

"At the end of winter," said Hubert, "large herds of seals come to whelp on the ice-floes that surround the archipelago, and the inhabitants kill great numbers of these animals with clubs and gaffs. Though there is no market for them, the people of the Islands make good use of them: with the tanned hides they make clothes and boots, the oil they use as fuel for their lamps, and they survive until summer off the meat . . ."

As for the names and origins of the Acadian families, Barthélémy preferred not to be too specific; he was not thoroughly familiar with the history of the dispersed nation and, besides, his information was only hearsay. Some families had already established squatters' rights there before the arrival of Gridley, he reported, and others had been brought by Gridley from Ile Saint-

Jean. If memory served him correctly, there had been Noëls, Thériaults, Boudreaus, and – oh yes, he'd almost forgotten, a Snault, a certain Louis Snault, nicknamed Arseneau. Or so he'd been told.

Le Grand Snault was caught. This man must be a relative. From that day on he became an ardent supporter of the emigration project.

Suddenly, the mountain of obstacles that had prevented him from supporting Abbé Allain's plan collapsed like a house of cards. Taking the oath of allegiance seemed a small price to pay for the joy of embracing this uncle or cousin who went by the name of Arseneau! And then, hadn't the priest insinuated that taking an oath of allegiance (*prêter serment*) did not commit one forever, since *prêter* (to lend) implied the intention one day to take back?

In the Magdalen Islands, he reasoned, no one would ever die of hunger: there were fish, walrus, and seals in sufficient quantities to allow one to get through the winter without making undue demands on the livestock. And a man ought to be able to cultivate a plot of arable land. If the inhabitants of Miquelon emigrated there in sufficient numbers, Colonel Gridley would have no choice but to accept them. The Acadians would also profit from the protection of Mgr. Hubert – that, too, was no small consideration.

All winter, they busily prepared for their departure, working in the utmost secrecy. Abbé Allain's nocturnal visits to the house of Thaddée Snault multiplied, and each time he met new men who nodded their heads in approval of his proposals. Soon, it became apparent that the entire Acadian colony of Miquelon would take part in the emigration; even those who looked upon it with a lack of enthusiasm didn't want to be left behind, for fear of reprisals.

They must take everything they could, said the priest, cattle, tools, fishing gear, furniture, food, dishes. Too bad for the ungodly French regime that would have to bear the loss! But at that point they had to intervene and temper his enthusiasm, for they had only a score of small fishing vessels in which to transport all this cargo, not to mention the 250 people to whom it belonged.

Spring arrived, and Abbé Allain began to pray for his east wind. On that day in mid-April, after saying mass, he had the unmistakable conviction that it wouldn't be long in coming, and he decided to go and take leave of his colleague, Lejamtel.

He was a child of the sea, as much at home on the water as on dry land. He had made the crossing to Saint-Pierre so many times that he knew the entire route by heart. He let the wind carry him out of the Bay of Miquelon, until he was in a straight line with Mont Calvaire and Cap du Soldat; then he tacked to starboard and set a direct course for Ile du Grand Colombier. He had to tack twice more between Ile Vainqueur and the formidable cliffs of Saint-Pierre; and then, arriving in port, he lowered the sails and glided up to the quay.

A man named Gudin happened to be on the wharf that day to seize the mooring rope. "So, *citoyen* Allain," he called down to the priest from above, "were the seas rough today?"

Never in his life had Jean-Baptiste Allain been addressed in such an impertinent fashion; he felt a mixture of embarrassment and horror that made his flesh crawl. Times were changing, there was no question about it. There had been a time when he had always been addressed as *"Mon Père,"* or *"Monsieur l'Abbé,"* or quite simply *"l'Abbé,"* which, despite its curtness, still had a cordial ring to it. But *this*! He found a certain comfort in the knowledge that he would soon be far from all such ignominious treatment.

Despite the uncertain political situation in France, Abbé Lejamtel remained hopeful, and his hopes were not shaken when his colleague announced that he was preparing to flee Miquelon with the entire Acadian community at the first sign of an east wind.

"Just between you and me," Lejamtel said, rising from the table after dinner, "there's a slight possibility that I might also take the Acadians of Saint-Pierre with me. I'm waiting for news to reach us with the next boat. But we're not going to settle in the Magdalen Islands. We're thinking of continuing on to Ile Madame and Arichat. It seems that's where the Acadian nation is regrouping. It's the seat of the archdiocese, you know."

The image of an Acadia being rebuilt on islands linked by the sea flashed into Abbé Allain's mind, but Lejamtel jolted him out of his reverie by taking him by the arm and leading him to the window. He pointed in the direction of Pointe Gallante, where the great ocean swells were breaking into sheaves of spray. "Well now, it looks like you're going to have your east wind, after all," he said simply.

The next day at dusk, the entire population of Miquelon was on shore for the exodus. Draped in shawls, clusters of wailing, sniffling children clinging to their skirts, the women rocked infants in their arms, their eyes puffy with sleep, while the men finished loading the boats. They hustled to and fro, from the houses to the boats, tugging calves, sheep and pigs, pushing wheelbarrows piled high with furniture, provisions and other odds and ends, which they stowed as best they could. The wind had risen, and several people wondered if it was wise to take to the sea with a storm brewing in the inky vault of the sky. "Hell lies in the east," they said in the jargon of seasoned sailors; but Abbé Allain, his cassock tucked up to his knees, wading barefoot in the icy water to help with the loading, calmed their fears with a sally of wit: "Yes, you're right, but then Eden must lie to the west. And that's where we're headed!"

When all was ready, he recited a short Te Deum on the beach and then took his place in the boat of le Grand Snault, where the latter's wife and six children, along with Jude and Constance Chiasson and their three children, were scattered amongst cattle, poultry, sacks of flour and grain, furniture, rolls of cord, fish nets, sheaves of hay, sailcloth, barrels of herring and salt pork, and a huge cask of fresh water.

With the wind up, they had to row the boats into open water past Cape Miquelon before they could raise the sails. Then the little flotilla began to move with the wind, the lanterns in the masts of the boats winking as they rose and fell in the great swells of the Atlantic. It was cold, and the passengers, huddled together beneath large sailcloth tarpaulins, were unable to sleep. Even the cattle didn't stop bleating and lowing all the way. The seas grew higher with each passing hour.

Le Grand Snault and the priest reefed the mainsail twice, but soon they were obliged to strike the main sail and to continue only under the jibs.

"If you think this is bad, wait till you see the squalls off the Anguille Mountains," said Thaddée to his shipmate. "Oh, you haven't seen anything yet!" And a moment later: "Whose damned idea was it to wait till the worst possible moment to set sail?"

But Abbé Allain remained imperturbable. He contemplated the spectacle of his faithful flock riding out the tempest with the moist eye of a Moses leading his people through the parted waters of the Red Sea. He believed in all this business, the rascal!

"God is with us," he said. "Be of good cheer, God is with us."

Le Grand Snault gazed at him, smiling slightly, not at all sure that he was in agreement, trying to locate in the fog of his memory the historical figure this man brought to mind. "The one who heard voices," he thought finally, putting his finger on it. "Joan. Little Joan of Lorraine. Joan of Arc."

It wasn't until a few days later that he realized just how accurate his comparison had been. The storm had abated, but navigation remained very perilous, for now they had to look out for the scattered ice-floes that floated in the current. In the early hours of the morning, the priest awakened Snault from the half-sleep into which he had finally drifted. All about the boat the waves were bristling with little black heads, big round curious eyes peering up at them, while a graceful choreography of flippers and backs rose and fell in the swirling waters. Not at all alarmed by the presence of the boats, thousands of seals whirled about them like little wooden horses on a carousel.

Then Abbé Allain told Thaddée to listen: above the whistling of the wind in the rigging, above the slapping of the waves against the hull, above the creaking of the ribs and the snapping of the sails. And it seemed to the Acadian that he, too, could hear a sort of vague babbling sound, made up of clucks, yelps, whistles, coos, punctuated at times with sudden loud barks and snarls. He gazed at the priest in astonishment.

"The seals," said Allain. "They're talking to each other. They're talking to us."

And, suddenly, Thaddée Snault felt a great wave of confidence rising within him.

VII
The End of the Walrus

The setting sun was caressing the waves when the Acadians perceived, slightly to starboard, the silhouette of Bird Rock, the eastern sentinel of the Magdalen Islands.

This flat little islet, scarcely two acres in size, of steep cliffs smeared with guano, has, since time immemorial, been the refuge of myriads of seabirds. Passing it, remaining well out to sea to avoid hidden reefs, the emigrants must have heard the loud echoes of the cackling of this multitude of aquatic fowl, a further sign – after the seals, the schools of porpoises and the gigantic schools of herring they had encountered on their route – of the prodigious fertility of the Gulf waters.

Oh, this land they now saw springing up on the horizon seemed so full of promise! By moonlight, they crossed the rough waters off East Cape, their boats suddenly tossed about in an eruption of little waves moving in all directions; then they skirted a long dune that was transformed suddenly into a series of high gaunt cliffs split with deep crevices. After passing a particularly sharp cape that ended in a crag pierced with a number of high arches, they entered a vast bay protected on the south by a hilly island that resembled a giant reptile sprawled out on the carpet of water.

They dropped anchor in a small, shallow cove flanked by two identical, formidable-looking capes, like the stone lions sometimes seen guarding the homes of the world's powerful and rich.

From the beach, with its clumps of marram grass stunted by the winter winds, the land rose in long, gentle, rolling ridges for a few miles. Here and there, in the clearings, were little houses and farm buildings. Then the terrain rose sharply to a high ridge of hills, their smooth, bald peaks like the tonsured skulls of monks.

This was nothing like Saint-Pierre and Miquelon – especially not

Saint-Pierre – and yet, in some strange way, the landscape of this little archipelago reminded the immigrants of the islands they had left behind them, nearly five hundred miles to the east. They had the vague impression of having discovered here a replica in miniature of their forsaken fief. There were dunes, smaller than those of Langlade, hills more thickly wooded and valleys trimmer than those of Miquelon; and, of course, there was the sea, eternally the same.

Tireless surveyors of the horizon, the inhabitants of Havre-aux-Maisons caught sight of the flotilla riding at anchor in the cove and went down to the shore to get a better view; though without revealing their presence, for they had reason to fear the strangers that the tides occasionally washed up on their shores.

Since the Treaty of Versailles had given the Americans the right to fish the teeming waters of the Gulf and the Newfoundland banks, the inhabitants of the Magdalen Islands had known no peace, for the fishermen had been quick to abuse this privilege. Each year, they were invaded by fishing fleets from Boston and Nantucket, which, not content merely to destroy their fishing gear and occupy their waters, organized raids against their homes, ransacking their provisions, even raping their women. Unarmed – Richard Gridley had seen to that, fearing they might one day turn on him – they were at the mercy of these savage attacks and lived in constant fear of invasion.

It was not until they distinguished, amongst the jumble of objects and animals piled high in the boats, the presence of women, children, and even a priest that they ventured into the open and launched a rowboat in their direction. Subject to the sense of inferiority characteristic of all conquered peoples, they began by addressing the newcomers in English – "Who are you?" "Where do you come from?" – but when they discovered that the strangers were French, and Acadians to boot, their transports of joy all but capsized their little boat.

After greeting the exiles, the Magdaleners gave a hasty account of the conditions of life in the archipelago, speaking in half-sentences, allusions, and short, broken phrases punctuated by long, troubled silences even more eloquent than the words themselves.

It soon became clear to the newcomers that life was far from a dream under the iron rule of Richard Gridley. They had been living in France when the principles of republican democracy had come into fashion, and though they did not officially endorse these principles, they were sufficiently acquainted with them to understand that the lives of the Magdaleners, as these islanders were already called, were no better than slaves'.

A hard, shrewd military man, Colonel Gridley ruled his little empire with an iron hand. The islanders had no liberties: they were obliged to work for the Bostonian and to turn over to him, as tenure for their lands, two-thirds of all their produce: fish, oil, cattle, poultry, vegetables, and grains grown on their little plots of land. The remaining third they had to sell to him at the price he fixed.

With this tiny revenue, they were then obliged to purchase from him, and him alone – under the threat of expulsion and other ignoble reprisals – all their basic requirements: flour, molasses, cloth, salt, fishing gear, even fire wood, which they were strictly forbidden to cut in the seigneurial forest without special permission. Their clothing – if you could call the miserable rags they wore beneath their sealskin tunics clothing – was made from the wool spun from the fleece of their few sheep.

Clearly, "poverty" was not the word for the state of abject destitution in which these people lived. Yet, despite their gaunt faces and dejected eyes, despite the fact that their children ran barefoot in the late April frost, generosity was central to their lives. They invited the newcomers to partake of their meagre meal of boiled seal and crow meat and their thin tea of spruce needles, and to share their straw pallets, offering to put them up until they were able to construct lodgings of their own.

Looking back upon the circumstances that surrounded the arrival of the Acadians of Miquelon on the Magdalen Islands – the disenchantment that must have set in between the moment they set eyes on what appeared from Bird Rock to be the promised land and the moment they learned of the misery and servitude that faced them under the feudal regime of the colonel – one cannot help wondering what prompted them to remain there, rather than continuing on their way to some other, more promising land: Gaspé or Quebec or Arichat. Was it their deep-seated optimism in the face of adversity or the darkest sort of fatalism? Was it the compassion they felt for those other destitute Acadians, or a conviction that together they might see their way through the tyranny? Was it the unanimous desire of the group to remain, or was the

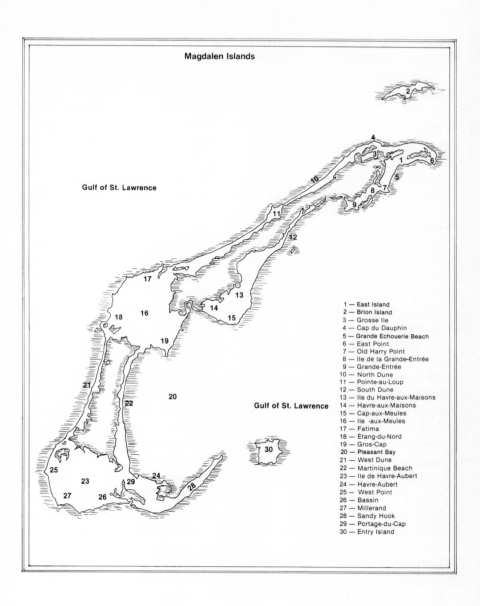

decision taken only by its leaders, Barthélémy Hubert, Abbé Allain, and Thaddée Snault?

Le Grand Snault had no sooner disembarked than he set off in search of his relative Louis, alias Arseneau, whom he finally tracked down in a small hovel deep in a valley surrounded by high buttes. He was an old man, more than eighty years of age, with a face as wrinkled as a dried apple, and he carried in the deep furrows of his brows marks of the calamities that had befallen him during the previous half-century. His small, bright blue eyes, hardened to sorrow and misfortune, betrayed not the slightest sign of emotion when Thaddée gripped him by the shoulders and took him in his arms.

It was not until Thaddée had given him a detailed account of his genealogy – *Thaddée à Charles, à Modeste, à Basile*[9] – that the old man's face lit up a little. At last, he opened his mouth, and in a thin, raspy voice that sounded like a rattle coming from deep in his chest, said, "I never thought I'd set eyes on any of you ever again."

Louis Snault, alias Arseneau, explained that he was the eldest son of Prudent à Basile, and consequently a first cousin of Thaddée's father; but what had befallen the rest of the large family, he did not know. Probably all dead. And his face clouded over as his lips pronounced strange, barbaric-sounding names: Cornwallis, Monckton, Boscawen, Wolfe – the very names of Acadia's executioners seemed to have predestined them for their horrible mission!

When their armies of ruffians had come to drive the Acadians out of Ile Saint-Jean, the old man explained, killing, plundering, sacking, he had been amongst those who took refuge in the woods with the Mic-Macs. They organized search parties to hunt them down, tracking them through the woods like wild beasts, even offering rewards for their scalps; and – sinister tribute to the friendship that had united the two peoples – the scalp of a white man was worth no more than that of his red brother.

Eight years he lived the nomadic life of the Indians, lurking in the depths of forests by day, crossing the arms of the sea by night in frail birch-bark canoes, racing down estuaries and little inlets in search of salmon and seal. He learned all the techniques for snaring animals and the technique of wielding a club to kill seal pups.

And then, peace was restored, or rather a pale semblance of

peace; for the vanquished people of New France were thereafter at the mercy of their victors. Back in Ile Saint-Jean, bursting with pride and triumph from the facile glory of the Battle of the Plains of Abraham, Gridley made it known that he would offer sanctuary to any Acadian who would agree to work for him on the Magdalen Islands.

Without quite knowing why, Louis Snault responded to this appeal and travelled to Souris to join the twenty or so families that the colonel's promises had lured out of the woods. But if he had had hopes of finding a better life in the Magdalen Islands, they were quickly dashed.

"Prisoners," he declared to Thaddée. "We were nothing but prisoners! Like animals, we were lured into a trap. And these islands where they brought us to hunt the sea-cow became our prison."

Thaddée and his companions, particularly Abbé Allain, were not yet ready to accept this harsh judgement; after all, hadn't they come to settle in the archipelago of their own free will? And yet, the leanness of the people, their appalling misery and their air of abject submission and impotence should surely have convinced them.

Several weeks after their arrival, when Gridley set them to work slaughtering sea-cows, there must have remained no doubt in their minds of the true nature of their condition: twenty hours a day they slaved, killing and sculping the beasts in an atmosphere of foetid grease and clotted blood, the odour of rotting meat rising constantly in their nostrils, plagued by big black flies and rats that swarmed to feast on the tripe. Stifled, nauseated, railed at by the foremen, forced to work for their very survival – because they had no other nourishment than this very meat they pawed over all day long – they, too, must have finally admitted that their lot was no better than that of convicts.

It must be said that, if times were difficult for the predators of the sea-cows, they were much more so for their prey. At the time of the arrival of the Acadians from Miquelon, these animals – commonly known as the walrus, or *morse* – had for thirty years been hunted indiscriminately by Gridley's men, as well as by droves of American hunters; and the species was already well on its way to extinction. Seven years later, the heightened competition having only accelerated the process of extermination, there would remain no further trace of the great herd off the Magdalen Islands – a resource once reputed to be inexhaustible, estimated at more than fifty thousand head. The last-known survivor of the herd was spotted off Ile du Bassin in the year 1800.

This is the heart-rending but classic tale of the encounter between the white man's mercantile civilization and nature. The moment he comes in contact with a natural resource, whether animal, vegetable or mineral, the white man feels compelled to exploit it in the name of progress, to bleed it dry with no concern whatever for the future. All that matters are the immediate profits to be gleaned, gold in the coffers, the blind, ephemeral prestige of the moment. Everything, even basic human dignity, is sacrificed on the altar of wealth and power. What does it matter if there should be nothing left for succeeding generations, if one day the oceans should be devoid of life, the forests stripped bare, the earth laid waste, and mankind the victim of a pestilential pollution that is the direct result of this enormous waste we call economy? Is there a single merchant or businessman on earth whose slogan is not, "After me, the deluge?"

Although this attitude has prevailed for many centuries, it is only today, on the eve of total ecological disaster, when even the rich and powerful are beginning to feel threatened by the scarcity of natural resources and by pollution, that feeble measures are being taken to try to stem the tide.

The encounter of the white man and the walrus dates back to the end of the ninth century, when a Norman navigator named Othère brought back several ivory tusks to King Alfred. Driven by covetousness, and perhaps by the belief that he was rendering a service to humanity (the walrus was then believed to be a sea monster), the king organized expeditions to wage war on this amphibious creature. Ships returned laden with precious ivory, which in those days was used as currency amongst certain Celtic, Tartar, and Mongolian tribes from the Atlantic to far beyond the Caucasus.

When it was later discovered that the fat of these pinnipeds yielded a precious oil, more valuable even than the ivory, the hunt was intensified. The Spanish, Basques, English, Dutch, French, and Danes vied with each other for this natural resource with such avidity that, in the space of a few centuries, the species was entirely exhausted. By the Renaissance, there remained practically no walruses in the north-east Atlantic south of the Arctic Circle, in the North Sea or in the Baltic Sea.

The hunters then turned their attention to the polar regions, particularly the areas around Franz-Joseph Land, Novaya Zemlya, and the Kara Sea, as well as to the New World, which had only recently been discovered by European explorers. There they found enormous herds of sea-cows in the Gulf of St. Lawrence and in the waters off Sable Island.

Although it had been hunted for nearly a thousand years, the walrus remained a mystery to the world of science until the great naturalist Buffon gave a more or less accurate description of the creature in his *Natural History* of 1765. Prior to that time it had been the subject of highly fantastic accounts, at times represented as a fish with horns, at times as a scaly monster with the head of a wild boar, webbed feet, and small curved fangs projecting over its lower jaw. One account described it as a hairy prehistoric saurian, ominous in appearance, with a barbed spine, clawed feet, and a fan-shaped tail. Another portrayed it as a unicorn decked out in an elegant suit of armour, like the knights seen on medieval tapestries.

The names given the animal were as varied as the descriptions of its appearance. Men of learning tended to refer to it as *vacca marina* (sea-cow), *bos marinus* (sea-bull) or *porcus monstruosus oceani germanici* (monster-pig of the Germanic Ocean); while, in the vernacular, it was more apt to be called sea-horse, sea-elephant or, more commonly, *bête-à-la-grande-dent* – large-fanged beast. The French name for the animal, *morse*, derives from a misconception on the part of a certain Olaus Magnus, the Bishop of Uppsala in Sweden, contained in a letter to the Pope near the end of the 16th century: "The coast of Norway," he wrote in his elegant script, "is inhabited by big elephantine fish called *morses*. They are so named because of their tendency to bite (*mordre*). When one of these creatures encounters a man on the shore, it attacks him, throws him off balance, tears him to pieces with its teeth, and kills him outright."[10]

All this was clearly very far from the truth: there are few pinnipeds as indolent and peaceful as the walrus – as long as it is not attacked, that is. Nevertheless the Bishop of Uppsala's blunder received papal sanction and was soon on its way to universal acceptance. At about the same time, a naturalist who had himself remarked that the tusks of this mammal emerged from the upper and not the lower jaw, concluded that the Creator had fashioned it thus to allow it to hang from rocks while sleeping!

We know today that walruses use their tusks to scrape at the bed of the ocean, unearthing the molluscs upon which they feed. We also know that they are gregarious, sedentary creatures, living for the most part in colonies, and that the walruses that once thrived in the waters of the Gulf and off Sable Island, and which probably moved south during the Ice Age, were identical to the ones that exist today only in the remote polar regions, where they are the object of strict protective measures. There are walruses in the Arctic Ocean from Spitzbergen to the Kara Sea, in the Bering Sea, in the Canadian Arctic, and on both coasts of Greenland.

Walruses are large, imposing, ungainly creatures, often reaching a length of three metres and weighing more than a tonne. They frequent shallow coastal waters and like to climb onto slabs of floating ice or crawl up onto beaches, where they lie packed tightly together. When they reach maturity – like the majority of pinnipeds, at about the age of six years – large fibrous tubers begin to emerge from their chests and shoulders, and it is these, along with their thick rolls of skin, their powerful neck muscles and their bristly moustaches, that lend them their highly uncomely appearance. The females, smaller than the males, nurse their young for two years, which partially accounts for the low reproduction rate of the species.

The first walrus hunters were amazed at the highly organized social and familial structures of the herd, and particularly at the ardour with which the creatures defended their young or any threatened member of the community. When one of their number was harpooned, it was not unusual to see them charge the boat and capsize it by hooking their tusks over the gunwales. In such instances, the hunters were in for a saltwater bath. At times – rare exceptions, to be sure, though they serve to demonstrate that these animals are not as dull-witted as they seem – walruses were known to push their thirst for vengeance even further: while the hunters were splashing about in the water trying to grasp hold of some piece of floating wreckage or to reach one of the other boats, they would turn their attention to the harpooner. The rowers and the helmsman were allowed to escape, but the man who had wounded one of their kin received no mercy. Together they attacked him and, with tusks and teeth, tore him to pieces.

However, it was only when they sensed they were being attacked that they became aggressive. As proof, let me recount the misadventure of the captain of a hunting expedition on Sable Island near the end of the seventeenth century.

The captain in question had accidentally fallen into the water while his vessel was anchored in the middle of a small herd of sea-cows and while he and his men were lowering the boats into the water for the hunt. Suddenly he felt two long, ivory tusks slide about his neck and drag him deep beneath the surface of the water. Down and down they went, diving straight toward the bottom of the ocean. Then, suddenly, his captor changed direction and carried him swiftly back to the surface of the water, depositing him more or less in the spot where it had picked him up. Apparently, he had been abducted by a female sea-cow who had mistaken him for her offspring, to whom she was about to give a lesson in underwater swimming!

In the Gulf of St. Lawrence, there were two main colonies of sea-cows, one with its breeding grounds on Ile Miscou in the Shippegan archipelago of New Brunswick, and the other in the Magdalen Islands. At the time of the discovery of the New World, both had been hunted by the native peoples from time immemorial, yet neither showed serious signs of depletion, for the Indians and Eskimos treated them with that same innate respect they showed for all natural resources, never killing more than they could actually consume.

Near the end of May each year, the Mic-Macs set up camp on the islands, patiently awaiting the invasion of the beaches by the walruses come to calve and mate. (Craftier than the Europeans, they preferred to attack these large beasts on land rather than on the water.) Soon the animals arrived: by hundreds and thousands, they piled up onto the beach, squeezing together almost to the point of suffocation, in a seething ungraceful flood that moved farther and farther inland as the newcomers pushed the early arrivals away from the shore.

When the entire herd was on land, the hunters availed themselves of a clear starry night and a favourable wind to practise what came to be known as a "cut."

With the wind blowing off the land, they approached by sea, so as not to arouse the animals' keen sense of smell; and, even before the old male standing guard could sound the alarm with loud trumpeting cries, they began to attack the beasts at the water's edge with pikes and clubs. There followed a formidable rout of panic-stricken walruses trying to reach the sea, their sole hope of salvation; but dogs specially trained for the task drove them toward the barriers created by the corpses of their slain fellows and, thus imprisoned, they became easy prey for the hunters.

After the "cut," the Indians left the walrus in peace. They spent several weeks carving up the carcasses, retaining only the meat, the hides, and sometimes a few ivory tusks, from which they carved tools, amulets, and other trinkets. All summer, they toiled at tanning the thick walrus hides, while the meat, cut into thin strips, was hung to cure in the acrid smoke of smoldering herbs and other aromatic bushes. Then, in the autumn, they made their way back into the interior with their booty.

The white man was intrigued by the ingenuity of the "cut" and, using his methodical mind, a truly formidable weapon, modified the tactic somewhat to serve his own insatiable appetites. With careful planning, a single "cut" could result in the slaughter of as many as fifteen hundred beasts.

Informed of the fabulous wealth represented by the sea-cows of the Gulf, King Louis XIII of France set up the Royal Company of Miscou just before the end of his reign. This company, in turn, founded the town of Nouvelle Rochelle, which was to prosper as long as the supply of sea-cows lasted, and which boasted at its peak nearly two thousand inhabitants, making it at the time one of the larger settlements in New France.

The sole industry of this town was the walrus. Relentlessly, the hunters killed and processed the beasts, dressed the hides, cut off the ivory tusks, melted down the fat, and quietly exploited the Mic-Macs as cheap, easy labour for the more odious tasks. During the summer and fall, trim frigates arrived to carry off the barrels of oil, sacks of ivory, and rolls of hides, heading back to France with their holds filled with the remnants of the untimely demise of a marine species.

Scarcely a century after it had been founded, the town of Nouvelle Rochelle had reverted to forest. All that remains today of this once-bustling industrial centre is a long artificial "beach" created by the bones of hundreds of thousands of slain beasts. As someone observed, the animals slaughtered there for the greater glory and wealth of the King of France have left a much more durable monument than have their assailants.

If the sea-cows of the Magdalen Islands did not meet with such a sudden and untimely end as those of Ile Miscou, it was undoubtedly because of the Basques. Hardy navigators, they were familiar with the archipelago long before Jacques Cartier officially discovered it during his first voyage to the New World in 1534. They were in the habit of wintering there while awaiting the arrival of the right whale, which they pursued throughout the summer as far inland as the mouth of the Saguenay River.

With their relatively mild climate, their well-protected harbours (Havre-aux-Basques, for instance), their forests abounding in game, their abundant supply of fresh water, the Islands were a particularly propitious place for the whalers to spend the winter. Not only did they lie directly on the route of the whales' migration, but the great seal herd that came there each spring to swarm on the beaches allowed the hunters to replenish their stores of fresh meat before undertaking a new whaling expedition.

The Basques considered they had certain inalienable rights to this territory, and each time French settlers attempted to colonize it, they sacked their settlements and forced them to turn back to Gaspé, thus preventing the establishment of a colony similar to Nouvelle Rochelle.

This is not to say that the sea-cows were not hunted on the Islands. On the contrary, they were the object of keen rivalry amongst a number of European nations, including the Basques. In 1597 this rivalry resulted in armed conflict between the English and the French, the first of a long series of skirmishes and battles between those two powers on the North-American continent.

In that year, when the British captain Charles Leigh arrived in the Magdalen Islands to join in the sea-cow hunt, he was astonished and angered to discover four other boats already anchored in the harbour of Grande Entrée: two from Saint-Malo and two from Saint-Jean-de-Luz in the Basque country. France and England were not then at war, but England and Spain were, and when Leigh asked the captains of the French vessels to turn their arms over to him as a matter of precaution, those from Saint-Malo concurred while those from the Basque country refused. The English were then obliged to seize them.

The Basques, who considered themselves at home on the Magdalen Islands, were not about to allow such a humiliation to go unavenged. The following day, they mobilized some three hundred Mic-Macs used as cheap labour on their killing grounds and, with the aid of the men from the Saint-Malo ships, led a victorious counter-attack, leaving Charles Leigh and his men no recourse but to flee. Thus, the first Franco-British conflict on the North American continent resulted in a French victory.

Of course, the English would have ample opportunity to erase that ignominious defeat. But, as the rivalry between the two powers intensified, the walrus colony of the Magdalen Islands profited from the hostilities; for while the men were occupied in waging war upon each other, they left the animals in peace and the herd had time to recoup some of its losses.

God knows the war lasted long enough, from the first movements of British troops during the War of the Spanish Succession (the capture of Port-Royal on September 24, 1710, sanctioned by the Treaty of Utrecht in 1713) right up to the decisive battles of Louisbourg and the Plains of Abraham half a century later. The walrus must have thought the good old days had returned! Only the Indians came from time to time to disturb their peace, but they were less savage and much less greedy than the white man, and the herd was easily able to make up the losses incurred during a century of intensive hunting. At the time of the arrival of Richard Gridley and his singular crew of "volunteers," it must have almost reached the size it had been 250 years earlier, when Jacques Cartier had explored the waters of the Gulf.

But the colonel was not a man to be moved by the fate of a marine species, particularly not with the Industrial Revolution sweeping steadily over Europe and increasing the demand for oil to an unprecedented level. The valves, pistons, and gears that, in greater and greater numbers, turned the wheels of the factories had to be lubricated. By the end of the eighteenth century, in London alone, nearly 200,000 tons of marine oil were being sold annually, and the tanned hides of the pinnipeds were being exported as far away as China.

No quarter for the walrus: that was the Bostonian's slogan. At two kegs of fine oil per adult beast, how could he let such a bargain slip through his fingers? The Acadians and the Mic-Macs who continued to frequent the Islands to draw upon this resource, were obliged to work like slaves on the beaches and in the killing grounds; and before long the sea-cows began to show signs of depletion, at first imperceptibly, then in a more dramatic fashion.

In 1783, the Americans came on the scene, and the hunt was

transformed into an orgy of extermination. The Mic-Macs, whose respect for all things in nature had been deeply offended, did not return to the archipelago. For the walrus, there would be no further respite: those that escaped the increasingly frequent "cuts" of Gridley and his men were almost certain to be harpooned by the flotilla of American fishermen awaiting them a short distance offshore, in the red froth of the waves. Male, female, young, old, all were slain indiscriminately, in the fear that any beast left alive would fall into the hands of the competitor.

Still, the animals would not admit defeat. They hung on for another decade or so, as if the formidable efforts of the hunters had merely intensified their instinct for survival. Wary of intruders, they kept far away from the boats and remained on shore only long enough to calve. While they were thus exposed to danger, they defended themselves against the attacks of the men and the dogs with a ferocity and ingenuity they had previously displayed only in the water.

The arrival of the Acadians from Miquelon was the final blow. Four years later, in 1797, it was all over. There remained so few sea-cows that it was no longer worth the effort to hunt them. The great herd that had once thrived off the Magdalen Islands had diminished so drastically that it was biologically beyond recovery. The few stragglers swam to and fro in the shallows, wondering perhaps why there was no one left to harrass them; they were destined for a death as peaceful as it was certain. The walruses, who had survived the moderate attacks of generations of predators on the shores of the Gulf, from the Dorsets and Beothuks, the Montagnais, Naskapis and Mic-Macs, to the first white adventurers, had found themselves sacrificed on the altar of the new-born industrial civilization, in order to lubricate the wheels of this gigantic system of exploitation and to line the pockets of a few merchants, the worthy successors of the feudal lords. Now Gridley could take his leave of the archipelago. He had no further business there; his fortune was made. What did he care of the future of those he had coerced into moving there, and who now found themselves with virtually no means of survival?

Today, another exploitable species, the tourist, has come to replace the walrus on the long, golden beaches of the Magdalen Islands. Indolent and tranquil, they lie packed together by the hundreds and thousands, to warm their hides in the sun. But I doubt that their disappearance, if that should ever come to pass, will leave as indelible a mark on the minds of the Magdaleners as did the demise of the sea-cow.

For it is not only in the bones and other remains that lie scattered about the sites of the former killing grounds, nor in a handful of geographical names – Seacow Rock off Ile du Bassin, Seacow Path, and, of course, Cape Gridley – that one can find reminders of those great marine mammals. Their memory is also firmly embedded in the subconscious minds of the Islanders. Long before men began to speak of ecology and conservation, the Magdaleners knew that no natural resource was inexhaustible, and that it was only by becoming masters of their trades and their lives that they would be able to maintain the abundant supplies of the marine species upon which they relied for their very ex'stence.

That lesson was taught them by Thaddée Snault, Barthélémy Hubert, Abbé Allain, and the other first-generation Magdaleners, who unwittingly and unwillingly participated in the extermination of the sea-cow. Of course, it would have been easy enough for them to exonerate themselves: they could simply say they had no choice. But the fact remains that, even as they toiled like slaves, they saw how little it takes to annihilate a resource that once seemed inexhaustible.

In the following years, when they set out for the ice each spring, they must have recalled the fate of the sea-cow and silently promised themselves that the harp seal would never be allowed to suffer the same fate, for it was now their only means of subsistence.

VIII
The Pagophiles

The seals existed long before man made his first appearance on the scene.

For millions of years, these amphibious creatures thrived in almost all the waters of the globe, at times gregarious, at times solitary, now sedentary, now on the move, dependent upon the sea that provided them with their food, but also taken to climbing out of the water, onto ice, rock, and sand, to rest, moult, and give birth to their young.

Their appearance as a distinct species dates back to the prehistoric Miocene Era, when the earth's crust was stabilizing after having been shaken up for thousands of years, and when the last saurians of the Jurassic contested the supremacy of the jungles and swamps with the great mastadons, the early ruminants, and the apes.

Their precise origins are unknown, though several theories have been advanced on the subject, the most popular being that they are descended from a terrestrial mammal, a prehistoric cousin of the cow or dog, which prudently returned to the water to escape the attacks of the big saurians. But they also share certain characteristics with penguins – the presence of a cloaca in the females of both species, and a similarity in their sounds – which lends support to the view that they may be related to birds; while their predilection for water has given rise to the supposition that they may be the descendants of the coelacanth, a species of fish that dates back to the Devonian era and that still survives today after more than three hundred million years.

There are about three dozen species of pinnipeds in the world, including the otary and the walrus. Some of these, like the sea lion (a fur-bearing otary) and the harp seal, still exist in large numbers; but others, like the monk seal, are on the verge of extinction. The reason for this can be traced directly to the fact that man – especially the white man of the Mediterranean and Nordic civilizations – has always depended heavily on these animals for their oil.

In the Mediterranean Basin, the cradle of our Judeo-Christian civilization, the seals were hunted intensively from the earliest times, by the Phoenicians as well as the Egyptians and Greeks.

Ancient Greek legends portrayed them as sly, wary creatures, shifty and elusive, symbols of a virginity attributable not so much to any superior will on their part as to a misanthropic unwillingness to mate. Mythology tells us that nymphs pursued by the gods transformed themselves into seals, and that Poseidon possessed great herds of them, whose protection he entrusted to Proteus. Judging from the very small numbers of these animals to be found in the Mediterranean Basin today, however, it would appear that this minor god acquitted himself rather badly of his task.

The pinnipeds of North America were more fortunate – until the arrival of the white man, that is.

Since the beginning of time, the harp seals have led a peaceful, nomadic life. True hobos of the sea, they cover tens of thousands of miles each year, swimming frantically from the icy boreal seas to the more temperate waters of the North Atlantic, remaining always on the edges of the ice-fields, stopping at either end of their journey only long enough to whelp and mate, and to gather strength for the long voyage back.

Nature and need have fashioned these creatures in such a way as to make them dependent upon the ice – not the hard, thick ice of the polar cap that several of their fellows (the ringed seal and bearded seal) thrive on, but the free-floating ice-floes, which offer solid platforms to climb on to warm themselves in the sun, as well as vast stretches of open water in which to frolic and sport.

Fossils discovered on the shores of New England indicate that harp seals, as we know them today, existed as early as the beginning of the Pleistocene Era, long before the advent of *homo sapiens*. Other fossils unearthed on the beaches of Virginia suggest that they moved southward during the last ice age.

In the summer, the pagophiles find their favourite habitat in the ice-choked waters of the Arctic regions. Returning from the south where they have whelped and mated, they break up into little schools along the west coast of Greenland, from Cape Farewell to the Lincoln Sea, and throughout the entire archipelago of the Canadian Arctic. There they frequent the sandy shallows and the mouths of fjords with their constantly melting glaciers, in search of polar cod. Great explorers, they are constantly on the move, and it isn't unusual to see large herds of them in Hudson Bay, in the coastal waters of the Belcher Islands, and even as far west as Churchill.

For them, this is a period of tranquility and rest. Food is abundant, and beneath the septentrional latitudes there are few predators – schools of grampus, the big orcs that have been wrongly baptized "killer whales," the occasional polar bear, and the aborigines; but their appetites are modest in comparison with those of the white man of the southern latitudes. While the adult seals go about their business, males and females strictly segregated, the young pups sport and frolic far from the watchful eyes of their elders.

The approach of the polar winter brings this vacation to an abrupt end. On September 21, the sun, which for six months has hung relentlessly on the edge of the horizon, illuminating with its pale light the spectacular landscapes of the high Arctic, drops beneath the equator and makes way for the long polar night. At the same time, the ice-cap once again begins to solidify. Group by group, the seals abandon the hospitable waters of Melville, Cornwallis, and Ellesmere Islands and the enchanting fjords of Umanak and Upernavik, those rich beds of cod and char where they have spent the summer feeding, and set out on the long trek southward. On the way, they encounter other groups of seals also fleeing the grip of the ice, and soon large herds of pagophiles begin to converge from all over Baffin Bay.

In the vicinity of Cape Dyer, on the Arctic Circle, where only two hundred miles of seawater – the Davis Strait – separates Greenland from Baffin Island, the herds from the western and eastern shores of the large Bay merge and continue their journey southward together, pursued by the tongue of ice that issues from the mouth of the strait and grows longer with each passing day.

At Cape Chidley, they join the seals that have summered in Hudson Bay and are fleeing it by way of Hudson Strait. Here, they rest for a few days, feeding and gathering their strength for the last stretch of the voyage along the coast of Cain Land (as Labrador used to be called), for they won't eat again until they reach the rich fishing beds of Newfoundland and the Gulf of St. Lawrence.

For several days, they devour large quantities of shrimp, plankton, algae, and sea cucumbers. They also swallow heavy stones, which apparently serve as ballast, allowing them to dive to great depths to fish the lower waters – unless, as some biologists believe, they indulge in this peculiar habit to avoid stomach cramps during their long periods of fasting.

Once they have eaten their fill, they set out again, as always in impeccable order. The old males lead the way, swimming abreast over a distance of several miles, attentive to the slightest sign of danger. They swim effortlessly, now on their bellies, now on their backs, sculling lazily with their rear flippers, using their front flippers only to change position or to alter their course. The pregnant females follow in their wake, first the older ones, then the younger; and behind them come the young males, ready to stave off an attack from the rear.

Less disciplined, the young form a separate group in which there reigns a joyful anarchy. They linger in coves, explore the depths of bays, come, go, twist, turn, giving themselves up to a whole variety of antics and comical escapades, which earned them the name by which the early French colonists knew them: *bêtes de la mer*. The English who succeeded them adopted the term, anglicizing it slightly to *bedlamers*.

These young seals are distinguishable from their elders by their size and their exuberant behaviour. Their coats, too, are distinctive – grey-beige and speckled with black spots. Each time they moult, the spots become larger, until finally they form a sort of crescent-shaped design on the back, sometimes resembling a heart, sometimes a saddle, sometimes a harp, which accounts for their name.

Thus, since time immemorial, the great nation of pagophiles can be found each November strung out along the barren coast of Labrador, moving with the cold currents in a southerly direction, famished and weary, but guided by a deep instinct that tells them to keep moving.

A French adventurer who settled in Cain Land at the beginning of the eighteenth century to take part in the fur trade saw them passing and marvelled at their great numbers: for ten days and nights, without interruption, the sea was full of seals, stretching from the shore as far out as the eye could see! Today, less than three hundred years later, all the harp seals of the northwest Atlantic can move past any given point on the coast of Labrador in a mere two days.

And yet it would be an exaggeration to say that the species is on the verge of extinction. Even the most pessimistic estimates set the present population of the herd at well over a million head.

Before the white man joined in the hunt, however, there were at least five times that number. But then there must also have been five times as many fish, five times as many cetaceans, five times as many marine creatures of all sorts preying on each other in the strict, complex biological cycle of the oceans. Man, of course, was not then the hypertrophied predator he was later to become.

Arriving on the scene too late, modern ecologists find themselves faced with a thorny problem, for if they allow a marine species to increase indiscriminately, they risk destroying forever the frail ecological balance upon which the survival of a large segment of humanity depends. So, rather than take such a risk, which might in the long run result in a dramatic increase in resources, they choose to cling to a policy that amounts to little more than maintaining the status quo of present exhausted resources, worked out in accordance with a handful of simplistic trophic bonds.[11] As their object is to maintain supplies at their current level, and allow for only a slight increase in the size of the herd, they pray with all their hearts that no new information will come to light to upset their precious theories!

When the seals reach the latitude of the Strait of Belle Isle, famished and exhausted from their long trek down the Labrador coast, they encounter the capelin that swarm in and out of the Gulf of St. Lawrence in large schools, moving with the rhythm of the tides, and they treat themselves to gargantuan meals of this small, spiny sardine.

What an abundance of food after the long period of fasting! There are so many of these little fish that the seals don't even have to exert themselves to feed: they simply open their jaws and swim against the current, chewing a little from time to time, crunching up the small fry that float into their mouths. When they've eaten their fill, they stretch out in the water, their bodies slightly inclined, only the tip of their snouts showing above water, and sleep off their meal.

At about the same time that these great feasts are taking place, another herd of seals arrives to join the pagophiles: the cystophores. A member of the pagophile family, this variety of seal also favours the ice and is also a gregarious, migratory creature; although within the herd it maintains rigorously independent family units. But here its kinship with the harp seal ends.

An enormous beast with a black head and a bluish hide covered with large black spots, the cystophore at maturity reaches the respectable length of two and a quarter metres and weighs up to half a tonne. It is, moreover, an ill-tempered creature, for unlike the harp seal, which is naturally gentle, sociable, and afraid of virtually nothing, the cystophore is known for its irascible, bellicose temperament. Rearing itself up on its fore flippers, it will stare an intruder straight in the eye with an air of such scorn and contempt as to strike fear into even the bravest of hearts. A man is wise to withdraw at such moments, for if the creature senses that it is being provoked, it will charge its adversary, throw him off balance and hurl itself on him, scratching and biting, and retreating only when the intruder is unconscious or dead.

This species of pinniped well deserves its name *cystophora cristata* or bearer of a crested hood, for the adult male has a sort of bladder or trunk on its head that descends from the top of the skull to just below the upper lip and that can be raised fully erect in moments of alarm or anger. Endowed with a number of inflatable appendages, the cystophore can also turn its entire nasal passage inside out in a paroxysm of anger, causing the mucous membrane to project rapidly from the nostril like a shiny red reed. This is an unmistakable signal of danger.

In the vernacular as well as in scientific circles, it is these characteristics that have inspired the great number of names by which this animal is known. Commonly called the hooded seal (in Newfoundland, simply "the hood"), in the other Maritime provinces it is referred to as the crested seal or the bladdernose; while, in the Magdalen Islands, it goes by the name of *"mâle de poche,"* or simply *"poche."*

A club is of virtually no use against these creatures. It will simply bounce off the inflated proboscis at twice the speed with which it was swung, and the hunter will be lucky not to dislocate his shoulder. Even against newborn pups, or bluebacks as they're called, the club is next to useless. They have to be killed from a distance, with rifles; and it's a sport in itself trying to recover the corpses of the slain pups, whose pelts are worth as much as $50, in the face of their enraged parents. Biologists who brand and tag these animals for future identification as part of their scientific studies have also learned to treat them with great respect: with the wisdom gleaned from experience, they administer a light sedative to the creature by means of a syringe mounted on the end of a long pole before proceeding with their work.

At the onset of winter, the herd of hooded seals, which even today numbers more than a hundred thousand beasts, joins the great herd of harp seals in the waters of the Strait of Belle Isle. But they do not mingle. The hoods, better divers than their smaller cousins, remain several miles farther out to sea, where the water is deeper: in the region of the Front, they can be found to the north and the east of the great herd; in the Gulf, to the south and the west.

If it's possible to speak of the harp seals as the hobos of the sea, the cystophores are its streamlined ocean liners. Each autumn, they instinctively assemble in the vicinity of Cape Farewell, at the southern tip of Greenland, and from there follow a direct course for the Strait of Belle Isle at a cruising speed of some fifteen knots. Fiercely independent and monogamous, individual couples often stray far from the herd. They have frequently been spotted as far north as Ellesmere Island and occasionally as far south as Florida. Once the whelping and mating seasons are over, they quickly forsake Canadian waters and set out once again for Greenland, whose shores they reach about the end of April. By the middle of the following month, they can be found in the Denmark Strait, between Greenland and Iceland, where they moult.

In all other respects, the behaviour of the cystophores resembles that of the pagophiles, though their activities are centred more on the family unit than the community. Bearing in mind that it is forbidden to hunt these animals in the interior of the Gulf (though, alas, not on the east coast of Newfoundland), and that the little bluebacks come into the world sporting their marvellous slate-blue hides, having shed their foetal fleece inside the womb, we can consider the two species together as part of the great pagophile nation, so as not to further complicate matters.

After feeding for two or three weeks on a nourishing diet of capelin and deep-sea fish, the pagophiles and cystophores are back at full strength. Then they break up into two herds. Governed as

Distribution and Whelping Grounds of Harp Seals in North-Eastern America

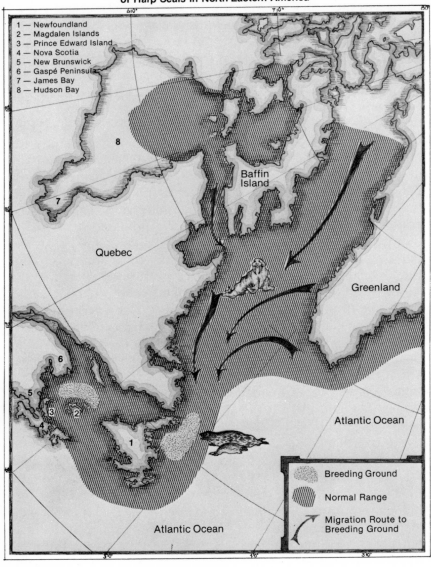

much by chance as by necessity, one of these moves into the interior of the Gulf while the other remains off the east coast of Newfoundland, in the area known as the Front.

In smaller groups, they spend the months of January and February exploring the rich fishing beds of the Gulf and the Grand Banks, feeding on the abundant supplies of fish, crustaceans, and molluscs, as they did in the high Arctic during the summer. In late February, at the appearance of the last winter moon, they assemble once again – those in the Gulf in the region of Bradelle Bank and Orphan Bank, and those in the Front, to the northeast of Belle Isle. Then the females climb onto the ice to whelp.

Why here rather than elsewhere? Presumably because the seals prefer hard, solid ice on which to give birth to their young, in order to ensure them a secure habitat. In the Gulf, the only ice-floe that meets these requirements is the one discharged by the St. Lawrence River; in the Atlantic, it is the thick, black ice of the polar cap, several years old, which happens to be in the region of the Front at this particular time of the year. The whelping ice is so distinctive that an experienced sealer can locate the main patch simply by studying the floes. In the old days, during the rare winters when there was no ice in the Gulf, the seals were known to climb onto the snowy beaches of the Magdalen Islands to drop their young.

The birth of the pups is a collective affair. The females gather in groups of four or five and begin by testing the thickness of the ice, an operation which involves raising themselves on their flippers and letting themselves fall onto the ice with all their strength. The next step is to prepare a nest in the shelter of the hummocks, preferably in a spot where there is a little soft snow. Then, if there is no lead in the immediate vicinity, they must dig a water hole.

Without tools, this is a task that calls for great ingenuity. Certain seals of the Antarctic literally bore a hole in the ice with their muzzles by spinning in the water like a drill; but the pagophiles operate in a less spectacular fashion. Like their cousins of the antipodes, the harp seals attack the ice from beneath, where its contact with the salt water makes it as soft as magma; but instead of drilling, they warm it with their breath while scraping it with the strong talons of their flippers. On the surface, the water hole resembles a small conical swelling, like a miniature volcano several centimetres in height; in the centre is a funnel-shaped chimney just large enough to allow the passage of a single seal.

The pregnant females could not survive without these water holes, and they zealously keep them open by taking turns plunging through them every two or three hours to break the skin of ice that is constantly forming over them.

The scope of the seals' invasion of the ice can only be seen from the air: the ice-floes, scattered over an area of some four hundred square miles, are literally black with seals. A day or two after they take to the ice, the females give birth to their young. The whole business is over with very quickly and is so well synchronized that, for a time, one has the impression that it is snowing seal pups.

First, the mother seal stretches out on the ice. She remains like that without moving, as if asleep, except that her eyes are wide open. Then, very slowly, she begins to twist, whimpering a little. Several minutes pass, and then she suddenly rises up on her front flippers, her body arched, in the grip of a series of powerful convulsions. Almost at once, the pup emerges from her cloaca and she severs the umbilical cord with her fangs.

Without stopping to tend to the little ball of life palpitating at her side, she moves at once to her water hole, leaving a trail of blood on the ice behind her, and disappears for several minutes. When she returns from her bath, she turns her attention to her pup, licking away the thin membrane covering its body and polishing its silky coat. Then she lies down on her side and offers it her breast.

The milk of the seal is extraordinarily rich. As it contains five times as much protein and fat as that of a human, the whitecoat grows at an astonishing speed. Small and puny at birth, its skin hanging from it like an oversized coat, its enormous, grotesque flippers resembling the wings of a gull, it is plump and fluffy after only three days of breast-feeding. By the fifth day, its weight has doubled. Two weeks later, when it is weaned, it has grown to a length of twenty-five centimetres and weighs nearly forty kilograms!

It is possible, however, that mother's milk is not solely responsible for this prodigious growth. Scientists have recently discovered that the foetal fleece of the whitecoat absorbs solar energy, particularly ultra-violet rays, and it may be that this energy is converted by the metabolism of these singular nurslings into fat.

During the early nursing period, the mother seal fasts, never leaving the side of her pup except for occasional quick baths. Night and day, she remains stretched out before it, attentive to its slightest needs, the very image of maternal affection. Neither hunger nor storm nor the occasional advances of the rutting males, who have begun to cavort in the waterways and who occasionally pluck up the courage to pursue their courting on the ice, causes her to ignore her duties as a mother. If a male approaches her and her little one more closely than she deems proper, she will throw herself on him with a roar and drive him back into the water.

The only thing that will cause her to flee is the arrival of a polar bear or a party of sealers. And even then, at times, her maternal instincts prove stronger than her desire for survival. Besides those exceptional mothers who defend their offspring with the ardour of despair, scornful of attacks and even of death, sealers have also reported seeing mother seals fleeing with their little ones, pushing them towards the nearest lead. The pup, who hasn't yet learned to swim and who is naturally terrified of the water, whimpers and wails; but, ignoring its protests, the mother rolls it into the water, plunges after it and, supporting it with her muzzle and flippers, propels it to the farther side of the lead. All the way across, the pup bellows and gasps, its lungs filled with the water it has inadvertently swallowed.

About ten days after birth, the whitecoats begin to moult. Two weeks later, when the shedding process is well advanced, the mother seals abandon them and turn their attention to the males.

This is the mating season, the moment the impatient males have long been awaiting! In the waterways, their performances have taken on an air of unrestrained revelry. By the dozens, they swim in circles, emitting long whistling cries and giving themselves up to all manner of acrobatic feats, in order to attract the attention of the females up on the ice. They leap and somersault, they swim on their backs, they twirl, they bark, they dive abruptly and come flying back up through the waves at a dizzying speed, in a great flurry of flippers and foaming water. From time to time, they gaze amorously at the females, then return to their aquatic displays.

Stretched out on the ice, the females contemplate this ballet, at first indifferently, then, as the males' capers become more extravagant, with greater interest. Some waddle along the edge of the lead to examine their suitors at closer range; they stretch their necks and make clucking sounds, as if to attract their attention; but the moment one of the males moves in their direction, they slip away quick as lightning. Other more flirtatious females actually take the initiative: they slide into the water and swim right up to the males who, maddened with desire, begin to whirl frantically about them. Feigning disdain, the females haughtily swim back to the edge of the stream and climb back onto the ice.

Then it is the males' turn to visit. A courting male will present himself before the female of his choice and make advances by beating his flippers and jigging wildly up and down. If the female does not consent to his advances, she will bristle, puff, bellow, and drive the intruder off with her flippers. If he persists, she may even bite and scratch him, throwing herself on him as ferociously as a tigress, letting up only when he has returned, humiliated and bleeding, to rejoin his fellows in the water.

If the suitor pleases her, she will likely accept his amorous advances and observe his performance with an air of indulgence, even pleasure. She will listen with delight to the raucous chant of his barks and respond to his courting gestures with appropriate whimpers and groans. Satisfied, the suitor will then waddle back to the water and give a new demonstration of his swimming skills in order to finalize his conquest.

These amorous games cause great jealousy amongst the males. Feelings of enmity break out and often result in fierce, bloody battles that end only when one of the adversaries is finally subdued, his neck torn and bleeding.

When the female is ready to couple, she returns to the water's edge and once again stretches out her neck. In the lead, the males suddenly try to outdo each other in barking. She advances, retreats with quick little bounds, then draws herself up with all her strength and wiggles her chest, executing little ballerina-like pirouettes, her body arched tight as a bow, only the tip of her navel touching the ice.

Her suitor then catapults onto the ice and wiggles about her, while she continues to dance her strange little hypnotic ballet. Brusquely, without the slightest sign of gallantry, he pushes her into the icy water, which bubbles and froths at the spot where the two mating seals have momentarily disappeared. They return to the surface a short while later, coupled tightly together, and for a while they float in the water like big black balloons, immobile, gasping, drunk with sensual bliss.

In this position, they are easy prey for the sealers who plough the waterways in their little boats, shooting at them with their 30-30s or their 303s. It often happens that when they manage to hook a slain male – it is always he who has the upper hand in these affairs – they find him still coupled to his mate.

Once coition is over, the seals climb onto the ice and stretch out side by side. But soon, they will return to the water to repeat the sexual act. If the female refuses to be mounted more than once (though this is more the exception than the rule), the male will dive back into the lead alone to seduce another female. After having coupled with her, he will lead her back onto the ice and leave her stretched out beside his first conquest. Thus, he may build himself up a little harem, and he will soon have to devote more effort to defending it from the other males than to enjoying himself. However, such harems are rare amongst the harp seals, for, unlike other pinnipeds, they are habitually monogamous.

While the adult seals – nearly two metres long, 150 kilograms of muscle and fat – are busy mating, surviving on love and pure water, as the saying goes (apparently they do not eat during this period), the pups finish shedding their white coats and, obedient to their gregarious instincts, assemble in small groups. They cross the vast ice-packs in single file and throw themselves one by one into the water, where they teach themselves to swim. Awkward and clumsy during their first dives, they soon master the technique, so that after only a few fumbling attempts they are quite at home in their new milieu, diving and swimming about with all the grace and agility of their elders. This is vital to their survival, for since weaning they have lived exclusively on their reserves of fat. Now, they must feed.

March gives way to April, and the cold, sunny days preceding the equinox are supplanted by days of rain and fog, sleet and hail, and the huge ice-field begins to break up, crumbling into smaller and smaller fragments. At sea, off the Magdalen Islands, reddish-tinted icebergs can be seen floating by in the currents, their strange colour due to their rough journey around the red sandstone capes of Prince Edward Island. In the region of the Front, too, the ice starts to disintegrate. The mating season is over. The pagophiles take to the water, where they feed on the shrimp and lobster that are slowly awakening from their long winter sleep in the rocky shallows. But soon they will return to the ice to shed their coats.

This time, they look absolutely exhausted; their fur hanging in long, loose folds from their bodies. They sprawl by the thousands, packed tightly together, lying in their own excrement and the thick tufts of hair they continually tear out with their claws in an effort to ease their terrible itching. They roll onto their backs and scratch their flanks with their flippers, emitting loud grunts and whimpers pitched to the level of their discomfort. Some males still carry on their necks marks of the deep wounds inflicted during the mating season. All in all, the ice-field has the atmosphere of a battlefield hospital, where the victims of gangrene slowly rot in the absence of proper hygiene.

After ten days or so, moulting is over and the seals dive back into the water to christen their new coats, which each year sport in a more dramatic fashion the characteristic horseshoe design that extends from the neck to the lower back. But they do not yet leave their breeding grounds in the Gulf and off the shores of Newfoundland, for the break-up of the ice invariably brings large schools of herring to spawn in the quiet coastal waters and the inland lagoons.

Once again, the seals gorge themselves, accumulating reserves of energy for the long journey back up the Labrador coast to the high Arctic, for this time they will be swimming against the current.

At the beginning of May, ignoring the schools of cod that have come in search of the herring, the harp seals degulf – as the Islanders put it – by way of Cabot Strait and the Strait of Belle Isle, on their way to rejoin the herd at the Front off the eastern coast of Newfoundland. An almost imperceptible increase in the temperature of the water has awakened their ancient nomadic instincts.

Little by little, the great herd reforms; the old males take the lead, breaking the waves with their hard-nosed snouts, followed by the females, and the young males bringing up the rear. The horde of bedlamers and beaters – those that have escaped the clubs of the hunters – form a separate school. And, one fine morning, the entire colony can be seen setting out on the long and perilous journey, swimming relentlessly towards the scattered ice-floes that lie somewhere far to the north, swimming toward the icy waters that teem with large cetaceans, cod and char, and with white belugas that will soon be giving birth to their young, by the tens of thousands, in the sandy coves of some lost little island, in the land of the midnight sun.

IX
Of Seals and Men

Since the beginning of time, the harp seals have travelled annually from the polar regions to the more temperate waters of the North Atlantic and back again, a great legion of life in pursuit of its mobile habitat, circumnavigating a course of some twelve thousand marine miles each year – eating, fasting, breeding, and dying of old age when fortunate enough not to fall prey to a polar bear, a grampus or a hungry walrus. The ocean they inhabit teems with life, from the *mysis* and *themisto*, those little shrimps of the boreal waters, through the entire gamut of clupeoids, scombroids, gadoids, cephalopods, and sharks, to the great cetaceans – not to mention the myriads of piscivorous birds, penguins, puffins, guillemots, cormorants and other wild fowl that inhabit the rocky islets offshore, and the big tortoises that were still fished in the Gulf of St. Lawrence less than a half-century ago. In shoals, schools, flocks, sometimes alone, this world of marine fauna is in a state of constant flux, driven by whim and need, in response to the urgent, albeit unconscious, struggle for survival.

The arrival of the first men, in the remote past of prehistory, did not drastically alter this picture of teeming abundance. For these were nomadic peoples and their numbers were few; their philosophy was "live and let live," and they took from nature no more than they needed for their survival. It was the intervention of the "civilized" white man, driven by excessive ambitions, that tipped the ecological scales in favour of impoverishment and propelled the entire ecosystem of the Gulf into a spiral of destruction and annihilation from which it may never recover. What the primitive peoples ("savages" and "barbarians," we call them!) did not accomplish in the five or ten thousand years of their existence on the shores of this great inland sea, the white man achieved in a mere four centuries.

A sad achievement, indeed!

Today, our poor Gulf of St. Lawrence is little more than a septic tank for some fifty million wasteful, over-nourished, super-industrialized North Americans, whose ecological consciousness is, on the whole, still embryonic and largely subordinate to their material well-being. It did not always sport the wretched, viscous face it wears today. The tales of the old folks, the accounts of the voyages of the early explorers, the nautical journal of Jacques Cartier himself, all paint such a different picture of the Gulf that it is difficult to believe they are referring to the same body of water.

Where are those pure, limpid, incomparably transparent waters through which one could see, several fathoms down, huge schools of cod, some of them as big as a man?

Today, they are thick and turbid, contaminated by the discharge of millions of toilets and sewers staked out along the vast drainage basin of the St. Lawrence: Duluth and Thunder Bay on Lake Superior; Chicago and Milwaukee on Lake Michigan; the mines of Sudbury and the petrochemical industries of Sarnia on Lake Huron; the metallurgical megalopolises of Detroit, Buffalo, and Cleveland on Lake Erie; the industrial centres of Hamilton, Toronto, Rochester, and Syracuse on Lake Ontario; the cities of Ottawa, Montreal, Trois-Rivières and Quebec – all spread out along this great open-air sewer whose sticky waters are not only thick with refuse, debris, and sedimentary rubbish, but poisoned by sulphuric acid rains, cyanide wastes, and the systematic discharge of mercury from the pulp and paper industry.

All this loathsome slime, mixed with the corrosive residues of the formidable industrial empire supported by this vast hydrographic system, is eventually disgorged by the tens of millions of tons each year into the Gulf, where it adheres to the banks of the Laurentian Channel whose waters move leisurely toward Cabot Strait, the sole access to the ocean depths.

Once, while working on a trawler, I happened to haul from a depth of some ninety fathoms a net full to the brim of this slimy, foetid substance that the fishermen call quite simply "a bag of shit." Though my stomach is strong and not readily subject to seasickness or the nausea often occasioned by the offensive effluvia of fish entrails fermenting in the sun, the putrid odour emanating from that pool of viscous mud actually made me vomit! The fish, crustaceans, molluscs, and benthonic fauna steer clear of this human refuse, with the result that they are often driven out into the ocean depths where they are more susceptible to attacks from their natural predators.

Added to this enormous sedimentary mass are the half-million tons of dust-like particles that fall annually from the atmosphere into the Gulf, the five million tons of telluric pollution that pour into it from its tributaries, and the accidental discharges of oil, motor fuels, and toxic chemical products that occur regularly along its route. To make matters worse, human genius has seen fit to block the Strait of Canso – the "gut," as it was commonly called – which used to siphon all this filth into the highly salted waters of the Gulf of Maine. From all this, it is easy to see that the role of "biological pump" which the Gulf of St. Lawrence is intended to play in the ecosystem of the continental plateau has been severely compromised. The dumping of hydrocarbons has resulted, moreover, in the formation of an oily film over the entire surface of the water, which reflects the sun's rays and greatly hinders the photosynthesis of the plankton.

Although many scientists believe that pollution, not sealing, is actually the greatest menace to the survival of the seal colonies, it is not the only agent responsible for the general impoverishment of marine resources, though it does dramatically slow the replenishment of existing stocks. These resources have also been drastically over-hunted and over-fished. However, like the degradation of the environment, the over-exploitation of the fishing banks and the wholesale destruction of entire herds and flocks of marine mammals and birds have sprung directly from the needs of our industrial civilization. Consequently, industry must be held largely responsible for the majority of ills inflicted upon nature.

I cannot help but be amused by certain ecological crusaders who, incapable of rising above the mentality of city-dwellers, see nature as a vast parkland that must be protected at all costs from the contemptible negligence of man and preserved in its virgin state simply to satisfy the voyeuristic tastes of vacationers – those faithful members of the industrial society who, for a few weeks each year, forsake their roles of unwitting/unwilling pollutors to feast upon the relics of all that was once beautiful and bountiful in this world. In their eagerness to achieve their sometimes laudable ends, these ecologists forget that there are other human beings (and these, not necessarily all aborigines) who still live off nature, and who have forged a role for themselves in the time-honoured drama of predator and prey. Rather than lauding these men as the precursors of a way of life that might one day lead the rest of mankind out of the merciless tyranny of cybernetics and machines, they berate them for various peccadilloes that irritate their super-civilized sentimentality, hoping in this way to bring them into step with the urbanized, industrialized, artificial and, forgive me for saying so, irremediably bankrupt system to which they themselves reluctantly belong.

I am, of course, alluding here to all those who, for fifteen years now, have been voicing indignation over the slaughter of "baby" seals. It isn't that I begrudge them their enterprise; it has achieved positive results, even for us, the inshore sealers of the Magdalen Islands. If the hunt had been allowed to continue unregulated, as was the case before 1969, there is no doubt that the harp seal would today be a threatened species, doomed as surely to extermination as the sea-cows of two centuries ago. With its formidable airborne arsenal, industry would not have hesitated to harvest the last survivor of the herd. But the Canadian government adopted very strict protective measures, and the size of the herd has dramatically increased during the past decade. I won't quibble with statistics, for I have a horror of figures; I shall simply say that, for several years now, the Gulf herd has not once failed to keep its rendezvous with the Magdalen Islands. Given the ballistic problem posed by the encounter of the herd (the projectile) with the Islands (the target) in the complicated play of winds and currents in the vast Gulf, this points indubitably to the fact that the herd has once again reached such a size that even the most unfavourable conditions make it virtually impossible for it to miss one end or the other of the archipelago that lies like a big hook in the middle of the Gulf.

Moreover, the experienced sealers – Wellie, Nazaire, Grand Jean – who are never wrong in these matters, have all remarked that the herd has undergone a revival. Needless to say, they rejoice at this fact.

A shadow hangs over their joy, however, for the Canadian government, succumbing to the mechanization that is quickly overtaking all activities in our society, and yielding to the pressures of the sealing industry, has recently decided, in view of the replenishment of resources, to increase the annual quotas allowed the big sealing vessels (although they were initially responsible for the depletion), to the clear detriment of the inshore sealers.

For government officials, small-scale sealing is "unaesthetic" in its anachronistic ways, covered as it is with sweat, blood, and blubber, while mass destruction of seals by means of machines evidently is not.

If only the whole sealing business could be mechanized, they think, then most of the criticisms and objections to the hunt would be automatically eliminated; for if there is one thing with which the abolitionists are in agreement, it is that the campaign is "unworthy of human beings."

This crushing argument – and it is only right to speak of it thus, in light of the club that crushes the skull of the unfortunate seal – is just one more example of the deluge of sophisms, half-truths and emotional pleas in which the dialectic of the abolitionists is couched. The campaign on behalf of the "baby" seals succeeded in stemming the holocaust of the pagophiles at the hands of big industry and made them the most highly protected species on earth, but since then all it has accomplished of a practical nature is to provide well-paid jobs for its star performers, thereby fighting unemployment.

It is not my intention to treat these crusaders as charlatans, for most of them, I know, act in good faith. However, some of these humanitarian organizations have built up prosperous concerns – judging, that is, by their ever-growing numbers and the competition amongst them for a corner on the compassion market. The main ones are the United Front for Animal Welfare, the Greenpeace Foundation, the CADAP (Comité d'Action pour la Défense des Animaux en Péril), the World Wildlife Fund, the OESAD (Organisme Européen de Secours aux Animaux en Détresse), the Foundation Franz Weber, and the European Committee for the Protection of Seals and Other Fur Animals. Organized like multi-national corporations, some have headquarters in several countries, collect hundreds of thousands of dollars annually and often possess their own fleets of helicopters or boats, as well as a vast array of highly sophisticated equipment with which they wage war on the hunt. The director of one of these organizations personally informed me that one of his rivals flies from capital to capital in his own private jet. Now, a Boeing costs in the neighbourhood of fourteen million dollars. A couple of years ago, the Canadian government had to withdraw non-profit status from the United Front for Animal Welfare, then based near Fredericton, New Brunswick, for excessively business-like attitudes. It moved its head office to the United States. So, whether it comes from the pockets of the great furriers or from those of their adversaries, a great deal of money changes hands because of the seal.

In the end, only the sealer emerges penniless, or with just enough money to keep himself and his family from starving. And yet, it is on his back that these colossal fortunes are built. The pelt, for which a sealer receives about fifteen dollars, doubles in value once the fat has been removed. Its price has doubled again when it returns from the Norwegian tanneries to be auctioned by the Hudson Bay Company in Montreal. By the time it has been made into a finished article, the fur brings about five dollars per square inch!

The motives of certain self-styled defenders of the harp seal seem contemptibly clear. Their methods – making super-realistic films filled with special effects that are often deliberately misleading,[12] to solicit funds that will allow them to stage publicity stunts and to make more films to solicit more funds – often put me in mind of a dog chasing its tail. This is not true, however, of all of them: most of the individuals and groups who oppose the seal hunt seem to be animated by sincerity, an authentic interest in ecology, and a genuine concern for endangered species. I am obliged to acknowledge this, although I maintain that their alarm is misplaced. On the one hand, the harp seal is no longer an endangered species; on the other, sealing, to the extent that it is not an appendage of the industrial monster, is a way of life for the local people. We are all too inclined these days to deprive men of their livelihood by replacing them with machines and forcing them to stagnate between the socio-economic crutches of unemployment insurance and social welfare, while the television plays full time to fill their empty hours.

This has been the painful tale of the Gulf fisheries for twenty years now. Formerly, the fishing fleet was made up of thousands of small crafts, each manned by a fisherman and his helper, who laid nets and lines and cast for fish with a jigger. Their total annual catch of cod, plaice, sole, hake, halibut, and haddock rarely amounted to more than 150 tons, thus ensuring the survival of both predator and prey. Then the factory-trawlers arrived from the old country and, with a creaking of pulleys, raised in a few trawls as many fish as the inshore fishermen could harvest in an entire season. To compete with them, the Canadian fishing concerns also outfitted their fleets with draggers, with the result that the fish, pursued into their spawning grounds and intercepted on their migratory routes, soon showed serious signs of depletion. They could not indefinitely sustain such a large-scale assault on their numbers.

The first to feel the effects were the inshore fishermen. They saw their catches reduced to a third, then a fifth, of their original size. In the end, it became impossible for them to make a living. Vanquished by machines, thwarted by technology, their only alternative was to seek social assistance to make ends meet. Today, the strict measures of protection adopted by the government have resulted in an increase of stocks; and the inshore fishermen might once again be able to ply their trade if, during the five years of their forced, subsidized inactivity, their equipment had not rusted and rotted. But in the absence of adequate assistance programs to help them get back on their feet (most government assistance goes, alas, to monopolies and big industry), they are reduced to following the daring feats of others via the magic medium of television.

Paradoxically, the apostles of ecology have not risen up in indignation at the over-exploitation of cod, the extermination of "baby" halibuts, the tremendous waste of non-commercial species by the big trawlers, and the disappearance of haddock in the Gulf waters; nor have they expressed shock at the dreadful agony of the lobster dropped into boiling water or the wholesale massacre of all those repugnant creatures – the jelly-fish, the moray eel, the octopus, the hog-fish, the toad-fish – which fail to strike a sympathetic chord in human hearts. For the herring that was once sucked up by the hundreds of millions of tons annually to make fertilizer and fish meal for livestock there were no protest marches, no petitions, no public burnings, no boycotts. And yet, this clupeoid, an essential link in the marine biological chain, came much closer to extinction than the harp seal.

The emotions aroused by an event like the annual slaughter of some 200,000 whitecoats (less than half the number born) are explained not only by the fact that the seal is a lovable, warm-blooded mammal endowed with a certain intelligence, and symbolizing in human eyes such qualities as restraint and timidity; but even more by the whitecoat's physiognomy, which dramatically excites the protective instincts of mankind, the same instincts that cause an adult to come to the defence of a child.

It was the celebrated ethologist, Konrad Lorenz, Nobel laureat in biology, who first opened my eyes to this fact. In his book, *Studies in Animal and Human Behavior*, he analyzes what he calls "innate releasing mechanisms," the physiological characteristics of a man or a beast that determine specific behavioural patterns in another man or beast. A female stickleback, for example, will react in much the same way to a red ball swinging to and fro in the water as she will to a male stickleback (which has a large red spot on its body) – by executing a nuptial dance. A male robin will furiously attack a tuft of russet-coloured feathers several square centimetres in size, as if it were a rival suitor.

Lorenz goes on to identify the characteristics which cause an adult to respond sympathetically to small children: a relatively large head; predominance of the brain capsule; large and low-lying eyes; bulging cheek region; short, thick extremities; a springy, elastic texture; and clumsy movements – the spitting image of the "baby" seal! He illustrates his theory with sketches of the head of a child and a man, the head of a jerboa and a hare, the head of a Pekinese dog and a hound, the head of a robin and a golden oriole. To his great astonishment, the reader discovers that he is inclined to react more indulgently towards the heads in the left-hand column than to those in the right – and all because of their physiological proportions!

Once it has been established that the whitecoat is a "releaser" of human protective behaviour, it becomes much easier to understand the magnitude of the crusade on behalf of the "baby" seals and the insane passions it often engenders. Once, during a demonstration to protest the war in Vietnam, I heard a member of an unlikely party of Quebec non-violent revolutionaries enjoining the crowd to forget about the atrocious bombings and the use of napalm on schoolchildren in that faraway country and to turn its wrath against the shameful, bloody slaughter of "baby" seals right here at home! Though the organizers of the protest march must later have regretted giving him the stand, there was nothing particularly astonishing in his aberrant view: the little, scrawny, rawboned Vietnamese is a much less effective "releaser" of human protective behaviour than the plump little whitecoat!

But then, you will say, if the indignation aroused by the mass slaying of whitecoats is phyletically normal, aren't the sealers the unnatural monsters they have always been suspected of being? And to that I reply: not at all, no more than those who slay lambs or sucking pigs (other "releasers") or the gluttons who consume them. Both groups have passed beyond the stage of emotive transfer. Moreover, seen from the lofty perspective of the sealer, the whitecoat, blending by mimesis with the ice upon which it lies, does not possess the same physiognomical characteristics as when photographed at eye level and displayed on the cover of *Paris-Match*!

The irony of the situation goes even deeper than that. If the sealers and the exploiters of the sealers – merchants, dealers, wholesalers, agents, shipowners, legal representatives, and magnates of the great oil and fur commerce – are in business, it's because there is a market for their products, and because this market has been created by all of us, by our very civilization.

Of course, it is always possible to boycott the fur trade, prohibit the importation of sealskins into the countries of the European Common Market, cast aspersions upon those who wear whitecoat hats or coats or mittens; but that does little more than attack the tip of the iceberg. Even a boycott of the leather wouldn't advance the cause of the seals one iota! It would still be necessary to do away with all the products in which seal oil is used; and that is virtually impossible, for marine oils, whether they be by-products of the walrus, the seal or the whale, have long been one of the staples of our so-called progressive civilization, used in such a wealth of commodities that I strongly doubt the majority of people would willingly do without. Rich in glycerine, fatty acids and alcohols, these oils are in demand for such a variety of products that it wouldn't surprise me a bit to learn that a few drops were involved in the fabrication of synthetic furs – a joke that went the rounds in the Magdalen Islands last year.

But, to be serious, what noble friend of nature would consent to do without his automobile simply because the mechanism of its gear-box is bathed in a solution of seal oil? And what self-styled ecologist would, for the same reason, boycott the bus, train, plane or even the helicopter that takes him out to the ice to frown upon the slaughter of whitecoats? And what starlet – whatever her feelings about the seal hunt – would be willing to give up her soaps,[13] her beauty creams, her depilatory lotions, her pomades, her perfumes, and her lipsticks, simply because the vast majority of beauty products contain some derivative of the fat of the beloved "baby" seals?

If she exists, she should first do away with chocolate, and a number of other candies; let her go and plead with the manufacturers of high-power explosives to abolish war, and with the revolutionaries to stop exploding their sticks of dynamite; for a glycerine extract from the fat of the whitecoats is consumed each time a powerful explosion is detonated! Then let her shut down all the chemical industrial plants, remove hundreds of pharmaceutical products from the shelves, prohibit the use of certain computors and space-exploration gadgets, and curb the indiscriminate use of a vast array of products available on the open market: television sets, photographic equipment, electrical household appliances, telephone dials, etc., for in the field of mechanics and cybernetics – especially in those regions of the world where temperatures reach extremes – there is hardly an axle, wheel, shaft or piston that slides, pivots, or couples, hardly a membrane that vibrates, without first being anointed with a greasy compound that contains at least a few drops of spermaceti or some other fine marine oil.

Ah, civilization! It is not only the marine mammals that have suffered from your repeated attacks; all the fauna of the Gulf have laboured for years beneath your yoke!

In this light, the journal of Jacques Cartier, which constitutes a sort of inventory of the biological assets of the vast region he explored prior to the destructive intervention of the white man, bears a strange resemblance to a necrological register, so numerous are the species to which it alludes that have utterly vanished from the shores of the Gulf. Moreover, the honorable navigator from Saint-Malo was, himself, the first intemperate predator of more than one species he encountered on his route.

After several weeks of ploughing the shifting ocean wastes, the *Grande Hermine* came in sight of Funk Island off the coast of Cape Bonavista, on the eastern tip of Newfoundland, and Cartier decided to put in for water. In his journal, he notes that this rocky islet was so covered with birds they seemed almost to be piled on top of one another, and he adds that there were a hundred times more in the surrounding waters and air. "Some of these birds," he writes in his old-fashioned orthography, "are as big as geese, are black and white, and have beaks like crows; they spend all their time in the water, being unable to fly in the air because of their small wings, which are more like half-wings, and with which they propel themselves as swiftly in the water as other birds do in the air. There are birds here so fat, it's a marvellous thing to behold."[14]

Having made this observation, he sent his men out to hunt these birds. In less than half an hour, they filled two boats with their prey, piling them one on top of the other like stones, "which, when salted, provided each of our vessels with four or five barrels of meat, not counting what we had eaten fresh."[15] And thus, the big penguins, which Cartier baptized *"apponatz,"* made their acquaintance with the voracity of civilized man.

Several weeks later, when the illustrious navigator reached the Bird Rocks (two islets in the Magdalen chain, one of which is now submerged), he found that they, too, were the habitat of large penguins, razor-billed auks, and solan geese. He supervised the killing of more than a thousand of these fowl by his crew, a task that took a mere hour and a half, states the proud captain, with his usual concern for statistical accuracy! Today, there is not a single penguin to be found either in the Gulf of St. Lawrence or on the shores of Newfoundland and Labrador – or anywhere else in the northern hemisphere, for that matter. It took only a few centuries for the white man to exterminate this species, which another people, the curious savages of whom I shall speak later in this chapter, had hunted for hundreds, even thousands of years without placing them in the slightest peril.

Like so many of their winged brothers – the passenger pigeon, for example, whose formidable flocks once darkened the skies of the prairies for days at a time and the last specimen of which died amid general consternation in the Chicago Zoo in 1934! – the penguins disappeared, victims of a civilization that placed a high premium not so much on their flesh, and their eggs as on their feathers, much in demand for European high fashion. How many other species of sea birds have vanished since the historic voyage of Jacques Cartier, I cannot say; but it is probable that a fair number had already met the same fate as the penguin and that others were quickly reaching that point when, in 1919, the Canadian government passed laws aimed at the protection of marine wild-fowl. Had they not been endangered species, it seems unlikely that anyone would have felt the need for legislation.

But let us return to our examination of the strange obituary-ledger of the discoverer of Canada. When he had salted his gargantuan supply of penguins, Cartier profited from favourable winds to proceed to Brion Island, where he saw immense herds of sea-cows, those "curious beasts, as big as full-grown oxen, with two tusks projecting from their jaws like elephants," whose sad fate has already been discussed. He also noted the presence of great numbers of bears, foxes and other furred and feathered creatures that are no longer to be found in those regions; they have disappeared along with the dense forests that once covered the Magdalen Islands, and that men neglected to replant as they exploited them. The trees and game proved as incapable of resisting the savage assaults of civilization as the fantastic schools of porpoises and dolphins that kept the *Grande Hermine* company on its voyage through Hoguendo Strait, the giant tortoises that Cartier observed in the waters of the estuary, and the white belugas that swarmed by the thousands in the mouth of the Saguenay River.

To this can be added the many species of fauna to which Jacques Cartier does not allude and which have now vanished from these regions – the colony of bearded seals that once thrived on the St. Paul Islands in the Cabot Strait, for example, and the giant squids, grey whales, and schools of haddock of the Gulf – though this doesn't even come close to completing the inventory of misdeeds perpetrated against nature by the white colonists. Last but not least of the vanished species are the aboriginal peoples who were either exterminated or cruelly decimated along with the beasts upon which they depended for their survival.

Of these, the Beothuks and the Dorsets merit a place in this book, for they were authentic peoples of the seal, the paleolithic ancestors of the present-day sealers. Their long residence on the eastern and western shores of Newfoundland and Quebec's North Shore testifies to the antiquity of the seal-oriented culture that still prevails in those regions, as well as in many other parts of the Gulf.

Before I turn to them, however, I would like to point out that all the peoples who have inhabited the Gulf region throughout history have been dependent upon the seals for their livelihood. The deplorably naked Mic-Macs that Jacques Cartier encountered in Baie des Chaleurs, for instance, and that so offended his modesty, were hunting and curing seals for their winter supply. Mic-Macs still exist in that region, but the white man's civilization has so corrupted their culture that today they no longer depend on marine mammals for their survival.

In Newfoundland, today, the Beothuk is widely exploited as a commercial symbol. In the Holiday Inn at Gander, for example, you can eat in the "Beothuk Dining Room," – a name calculated to spoil the appetite of anyone familiar with the barbaric manner in which this primitive people was systematically exterminated!

In his voluminous work, *A Journal of Transactions & Events on the Coast of Labrador*, the British Captain George Cartwright, who took part in the fur trade in that region in the last third of the eighteenth century, more than once deplores the genocide of these people at the hands of the Spanish, Portuguese, French, and especially the English, colonists of Newfoundland; and he prophesies their imminent end: "Their number must decrease annually, because our people murder all they can, destroy their stock of provisions, canoes and implements; these losses have occasioned whole families to die by famine." And, a few years later: "I am sorry to say that British colonists are much greater savages than the Indians themselves, for they seldom fail to shoot the poor creatures whenever they can and after it boast of it as a meritous action. With horror, I have heard several declare that they would rather kill an Indian than a deer . . . If the invasion of their territory by European settlers keeps on, there will not be any of them left before long."

His predictions proved to be exact: the last of the Beothuks died in 1829. All that remains today of that great aboriginal nation which once occupied all of Newfoundland except the Avalon Peninsula, are a few sketches of two prisoners taken to St. John's in 1813 to pose for posterity before being put to death, as well as a skeleton that is the pride of the museum in which it is displayed.

From these few clues, a great many theories have understandably been formulated about the history and the origins of the Beothuks. Although most anthropologists place them amongst the Amerindians of the "archaic boreal" genus, there are others who maintain that they are not Amerindians at all, but the descendants of a Western European tribe that probably immigrated to America at the beginning of the Quaternary.

There is no lack of evidence to support the latter theory. First, the Beothuks were tall (nearly six feet in height) and had light skin and eyes and auburn-coloured hair; they were called "redskins" because they painted their skin and hair with a mixture of red ochre and seal oil. When, in 1498, John Cabot brought eight of these natives back to the court of Henry VII of England, the chroniclers of the day noted, not without a trace of astonishment, that the "savages," when outfitted in English garb, resembled any other courtiers.

Their culture differed, moreover, in several respects from that of the other Amerindian nations: they did not keep dogs; they inhabited wigwams only in the summer, spending the winter months in log houses caulked with moss and capped with pyramidal roofs; and though they were not familiar with the magnetic compass, they navigated the high seas, ploughing the ocean waves in their big canoes that could hold up to thirty paddlers, sometimes travelling far from the sight of land in search of the small rocky islets that abounded with penguins and seals. This fact is of particular interest, for no other Amerindian people ever ventured onto the high seas. If the Mic-Macs of Cape Breton occasionally travelled to the Magdalen Islands, it was because on a clear day they could see the glitter of the archipelago's sandy beaches from the tops of their highlands.

Whatever their origins, the Beothuk economy was mixed, based in winter on caribou and, in summer, on seals and sea fowl. Each spring, after the snows had melted, they would emerge from the depths of the forests and settle on the shores of the vast bays along the east coast of Newfoundland. Their diet consisted of seal meat and penguin eggs. From the flesh, fat, and eggs of the pinnipeds and wild fowl they concocted a thick sausage, consisting of a coarse mash stuffed into the intestines of the seal, which could be preserved for long periods of time. Great quantities of this staple food were prepared for the following winter. Lacking spices, the dish must have been anything but a gastronomic delight, yet something about it seems to have tickled the European palate, for the Newfoundland colonists assassinated entire tribes of Beothuks for the sole purpose of appropriating their supplies of meat.

Thus, the seal, an essential staple in the subsistence of this primitive people of whom so very little is known, also became one of the causes of its extermination.

The Dorsets, on the other hand, never came into contact with European colonists of the post-Columbian strain. They disappeared from the face of the earth several centuries before the Renaissance. Yet their memories of the white man, with his blue eyes and his long shaggy locks and beards, must have haunted them like evil omens to the day of their deaths; for their sole brief encounter with him – in the guise of the Vikings who explored Newfoundland before the end of the first millennium – resulted in bloody confrontations, a premonition of the massacres and genocidal assaults that would later mark the encounter of iron-age man and his stone-age ancestors on the North-American continent.

Hailing from Siberia, the Dorsets were a race of giant Eskimos, often more than six feet tall, who spread throughout the Canadian Arctic and Greenland several thousand years before the beginning of the Christian era. They are widely believed to have emigrated southward at the time of what climatologists call "the little ice age" (approximately 3,000-300 BC), settling in Labrador, along the north shore of the St. Lawrence, on Anticosti Island, and in Newfoundland, where the prevailing climate would have been favourable for their culture and economy based almost exclusively on the harp seal and other marine mammals.

The Dorsets ate the meat of the seal, they used its oil for light and heat, and they made their clothing from its skins.

Since the early 1960s, Dr. Elmer Harp, professor of archaeology at Dartmouth College, has excavated and studied several ancient Dorset settlements in Newfoundland, notably those at Pointe Riche and Port-au-Choix, on the west coast of the Great Northern Peninsula. In each of the partially excavated houses, originally covered with skins stretched over a framework of bones, he found a quantity of tools and utensils that throw light upon the vocations of this primitive people: harpoons made of finely chiselled bits of flint and cleverly arranged bone ends, stone knives with formidable blades, hide scrapers, awls, and bone needles, as well as a number of amulets carved with effigies of seals and walruses.

Nearby, he also uncovered immense ossuaries containing the skeletons of innumerable seals. Their study revealed that the Dorsets hunted seals of all ages, including "baby" seals, which they must have killed in the manner of present-day sealers, with the aid of a club or a wooden cudgel, though these weapons have not survived the ravages of time.

The location of the living sites, augmented by the testimony of the Scandinavian sagas of the voyage of Thorfinn Kerlsefni to Newfoundland, has given rise to the theory that the Dorsets hunted a herd of pagophiles that whelped off the west coast of Newfoundland but that is now all but extinct. According to this theory, when the prevailing northwesterly winds shoved the icepack up against the coast, the Dorset sealers made their way out onto it on foot. Later in the season, when the ice began to break up and crumble, they hunted the seal from skin *canots* similar to the kayaks of present-day Eskimos.

And they survived as long as the seals lasted.

At the beginning of the Christian era, the climate began to warm, at first imperceptibly, then in a more and more dramatic fashion, until, about the year 1,000 the northern part of Newfoundland enjoyed temperatures of a near-Mediterranean range. Plane trees and wild vines grew there in such abundance that the Vikings named this newly discovered country, Vineland. With the disappearance of the ice from the Strait of Belle Isle, the harp and hooded seals forsook their old breeding grounds, causing consternation, anxiety, and famine amongst their predators. Each year, the great herds of the Front and the Gulf remained farther north and farther out to sea to breed.

Unable to adjust to a new way of life, the Dorsets had no choice but to climb back toward the boreal latitudes, their numbers dwindling with each stage of their retreat. All trace of them is lost somewhere in Labrador about the beginning of the twelfth century. With no fuss they vanished, incapable of adapting to the new conditions of life imposed upon them by the severe climatic changes. When, several centuries later, the temperature began to cool again and the pagophiles returned with the ice to the coasts of Newfoundland and the Magdalen Islands, the Dorsets were no longer there to hunt them.

But, by then, other men had come to replace them.

X
The Great Hunts

The French, it is said, were the first white men to hunt the harp seal in North America. Perhaps as long as a quarter-century before Jacques Cartier's first voyage to the New World, intrepid Breton fishermen had established fishing posts in the vicinity of the Strait of Belle Isle: at Karpon (now known as Quirpon), on the east coast of Newfoundland's Great Northern Peninsula; and at Blanc-Sablon and Brest (the present Bay of Good Hope), on Quebec's North Shore. In these capelin-rich waters, they fished for cod, which they salted and dried on fish flakes, trellis-like tables which were set out in the sun. Then, at the onset of winter each year, they turned their attention to another predator of the small, fleshy, spiny sardine: the harp seal. Arriving from the high latitudes of Labrador, the famished pagophiles invariably became entangled in their fishing gear and left their nets in shreds.

The dauntless Bretons must have felt little sympathy for this great horde of vandals and probably sought every means at their disposal to curb their devastating attacks. With stout line, they wove large, wide-mesh nets, several hundred metres long and several fathoms deep, and laid them in such a way as to protect their fishing gear. The seals, intent only on pillaging the lines, became entangled in the nets. All their struggles to escape only imprisoned them further, and they ended up drowning. Each morning, before going to raise their cod nets, the fishermen would haul in these seal nets and remove the corpses with which they were habitually filled.

It wasn't long before they realized that they could kill two birds with one stone: the seal, that noxious pest, could be turned to a profit instead of simply being thrown back into the sea. What a pity to cast such valuable furs, such fine blubber, and such good meat back into the lap of Neptune!

Gourmets and gourmands, like most Frenchmen, these sixteenth century Bretons undoubtably found the flesh of the pagophiles very palatable after the corned meat that was the staple diet of most crews. They seem to have been particularly fond of the brains and the sweetbreads of the bedlamers, as fine a delicacy as those of a calf! But these were men of the sea, whose occupations did not allow them the leisure to nibble at dainty dishes, and no doubt they were more interested in the generous layer of fat covering the body of the seal than in the meat. Melted down and poured into barrels, it fetched a good price in Saint-Malo, as did the *drâche*, the rancid cod liver oil they already brought back from their long voyages overseas. It is doubtful that they placed much value on the furs, for the preparation of the hides was a long and painstaking process, and the low prices they would have fetched in France probably would not have justified such an expenditure of time and energy.

Meanwhile, the oil alone made fishing for seals a lucrative business. All the adventurers who frequented the regions of the Strait of Belle Isle and the shores of Labrador throughout the centuries engaged in it. Even today, the same wide-mesh nets devised by the Bretons are used to fish harp seals during their long southerly migrations down the coast of Cain Land.

Along the lower North Shore and on certain islands off the coast of Newfoundland, where the seals do not simply sweep past, but stop to stalk the shallow coastal waters, gorging on the capelin that abound there, they were caught in traps, a method that is still practised today in the little Quebec and Labrador villages in the region of Blanc-Sablon.

The technique borrows elements from both hunting and fishing. A long net is stretched perpendicular to the shore, one end of it firmly attached to the beach and the other anchored at sea. Then another sunken net is laid from the anchor back to the beach, in such a way as to form a right-angled triangle with the first net and the shore. This is the trap or snare.

Several men are stationed on shore, while the others move out to sea in small boats and quietly encircle the seals. At a given signal, the men in the boats begin to make a great uproar, shouting, shooting rifles, ringing bells, blowing whistles, and they close in on the frightened beasts, driving them toward the trap. When the pagophiles have crossed the open net, all the men on shore have to do is pull on the bolt-rope to bring the submerged net rising to the surface of the water, thus imprisoning the seals. Up to their waists in water, the men then kill them with their clubs and rifles.

The practice of harvesting whitecoats on the ice-floes by the colonists of Newfoundland's eastern shores is intimately tied in with the climatic changes that have affected the North Atlantic over the centuries.

After the long warming period that lasted from the beginning of the Christian era to the eleventh century and that resulted in the extinction of the Dorsets, temperatures began once again to cool. Frequent violent cyclonic disturbances and a lowering of water temperatures a few degrees Celsius characterized this era, which continued until the middle of the sixteenth century. Then there was another warming period, which reached its peak a century and a half later and from which the European powers profited to establish colonies in Newfoundland, Acadia, New France, and New England. It was during this era that the practice of seal fishing was inaugurated: the ice had moved far out from the shores of Newfoundland, and the colonists evinced no desire to go and explore that dangerous and (as they believed) barren universe.

During the second quarter of the eighteenth century, the climate began to cool again, a trend which continued until the very end of the following century (since then, climatologists tell us, temperatures have once again been rising), and the ice moved in closer to the rocky shores of Newfoundland, where, each spring, it was literally covered with whitecoats. With the aid of strong northeasterly winds, the great herd of the Front unfurled like a tidal wave along all the Newfoundland seaboard. On such occasions, the entire population of the little fishing outports lent a hand in harvesting the seals: men, women, children, sometimes even the dogs!

Thus, the Newfoundlanders got into the habit of hunting "baby" seals. When the winds were less favourable, they would make their way out onto the ice on foot or thread their way through the leads in light little skiffs in search of the herd. In time, the seal became central to their economy: they ate its flesh, they made clothing of its skins, and they used its oil to provide their homes with light. Any surplus oil could always be sold to the merchants.

They didn't make any great fortunes however. The merchants always managed to find some excuse – the slackness of the market, the mediocrity of the product – to pay them as little as possible for their produce, leading them to understand that it was only as a personal favour that they consented to buy the "worthless" oil at all. But the little revenue the settlers did receive allowed them to escape the economic servitude to which they would otherwise have been subject. Several debts were wiped from the ledger – in itself a step in the right direction.

The merchants were lords and masters of the little Newfoundland fishing outports. They exercised absolute control over the fishermen, buying their produce for next to nothing and selling them their equipment and provisions at outrageously high prices, so as to keep them perpetually in debt. In this way, they obliged them to do business with them and them alone. Doubtless, they did not look favourably upon the practice of sealing, at least not in the beginning, for it escaped their control and somewhat undermined their authority over the fishermen. But business being business, they consoled themselves for their wounded pride and their unassuaged thirst for absolute power by making enormous profits in the commerce of seal oil.

Meanwhile, sealing became an increasingly adventurous calling, and the people of the northern bays, latter-day Vikings that they were, acquired a strange taste for it. Though in the beginning they had not ventured far from shore in their little twenty-foot boats, they soon began to extend their field of action to regions far out to sea. It became a highly competitive sport for those rugged seafaring men to venture out onto the huge polar ice-pack – twenty, sometimes fifty miles from shore – in light, open skiffs that offered no shelter from the elements, remaining for days or weeks at a time in search of the seals. Nothing daunted them: cold, wet, discomfort, blasting winds, drifting snow, freezing rain, blizzards – all were accepted as part of the game.

Just thinking about it sends shivers down the spine!

Of course, the swiling game – as it is still called – took the life of more than one inhabitant of those little outports, but this seemed merely to whet the competitive ardour of the others. Not even death was too high a stake for the chance to escape the tyrannical grip of the merchants; and each year more and more Newfoundlanders took part in the sealing campaign.

In 1804, 40 thirty-ton schooners and 50 open boats, manned by a total of 1,639 men, took part in the hunt. Those who made it back after the terrible storm that raged for nearly a week while the entire flotilla was out in the ice, returned with an unprecedented 81,000 hides; the others left theirs out on the ice. Of the 35 open boats that had set sail from the little harbours of Conception Bay that year, only ten returned; the remaining 25 were lost, together with the more than 150 sealers who had manned them.

This was the first great catastrophe of the seal hunt. All the little fishing outports on Newfoundland's eastern shores suddenly found themselves in mourning, a bereavement that was all the more cruel in that the families who had lost loved ones were now deprived of the very men upon whom they depended for their survival. For many, there would follow hardship, misery, famine.

Little by little, reason took hold: it was clearly foolish to confront the ice, far out at sea, in nothing but a little open skiff. The national sport of Newfoundland underwent its first major transformation – thereafter, only schooners would make their way out to the ice.

These boats had no heat, but seal fever more than compensated for this lack. As many as sixteen men would pile into one of these little thirty-foot vessels, their hulls sheathed with greenheart wood to protect them from the grating ice; and they would luff furiously out to the floes – the great black ice of the polar cap disgorged by Baffin Bay that spreads quickly with the currents throughout the North Atlantic.

Their great enemy was the east wind. When the boats were small, it had been a simple matter to lift them onto a solid ice-pan and build a shelter with them when it began to blow, piling the ice-fields up against the rocky coasts of Newfoundland and Labrador. But now this manoeuvre was no longer possible: wedged between the cliffs and the ice-floes, the skippers had no choice but to haul close to the wind and try to find refuge somewhere in the interior of the living ice.

They had to tack through narrow leads to the enclosed ponds that provided temporary havens, always at the risk of being crushed between two floes or remaining imprisoned in their provisional harbours. They had to navigate in all sorts of brine: thick slob ice, in which they often became mired; crystalline frosy (a thin layer of ice, sharp as a blade, that covers the surface of the water on cold, calm nights); leads filled with slabs and discs of ice that shot by like torpedoes in the current; and the great ocean swell studded with icebergs and all sorts of icy debris.

The fact that no yachtsman, however fanatical about sailing and the sea, has never dreamed of retracing their steps, is a clear indication that these were far from pleasure outings! And yet, the Newfoundlanders found a certain inexplicable charm in them, which the thirst for profits alone cannot explain. These were men of the ice, a breed of Vikings the like of which the world has seldom seen! Dressed in sealskins, they possessed, like the seals themselves, a familiarity with their milieu, a kinship with the ice-floe that allowed them at any moment to discern its smallest flaw, or anticipate its slightest whim, and thus to extricate themselves from its murderous grip. Their affinity for the ice was such that they carried no reserves of fresh water with them on their long excursions; like the seals, they melted ice to quench their thirst.

The look-out man, perched in the crow's nest at the top of the mainmast, was the one upon whom the safety of the entire crew depended. In the labyrinth of channels and leads that zig-zagged all the way to the horizon, it was he who chose the route that led to the seals. His decisions could be overruled only by the captain.

From the moment of the boat's departure to the moment it sailed back into port, someone was always on watch in that little tub that swung fifty feet in the air. In thick fog, in blinding blizzards or on the open sea, the look-out might be manned by one of the younger men, but when it was a matter of navigating through the ice to locate the seals, only the most experienced sealers were allowed to occupy the crow's nest; and when, between streams of tobacco juice, they yelled in their raucous voices: "Swiles! Swiles har' to starboard, sir!" their cries were as sweet to those down below on the bridge as the sounds of a celestial choir. But when, at the height of a storm, they called out an alarm, it was a mournful wail, frightfully human, that reached the men's ears through the whistling of the wind.

Nothing fazed these sealers. At times, to avoid being crushed, the captain had no choice but to run his vessel up onto an ice-floe lying flush with the water. All sails trimmed, he drove the boat head-first onto the ice in an attempt to raise the prow onto it.

Then, with half the crew harnessed to hawsers and the other half wielding thick planks as levers, the hull was lifted clear of the water.

Though safe, the boat was now a prisoner of the floe, and the entire crew had to go to work with saws and gaffs to cut a channel in the thick ice, at times over great distances, to the nearest navigable waterway.

If they became entangled in slob ice, they would have to resort to a technique familiar to whalers and walrus hunters in the boreal seas, commonly known as "rallying." The entire crew would be called up on deck, where they would fall into ranks on the larboard side. At a given signal, they would all race for the starboard side; at the next signal, they would race back to the larboard; and so on, to and fro, from one side of the deck to the other, running in unison, so as to impart to the vessel a pendulum-like movement, just as modern ice-breakers do by transferring their ballast from side to side with the aid of powerful pumps. If there was no wind, it would often be necessary to "rally" for hours to extricate the boat from the soupy brine; and if this didn't work, the men would be ordered out onto the ice, where they would attach themselves to hawsers and pull – like Volga boatmen!

But the worst hardships were caused by the raging easterly storms that came up unexpectedly and unleashed the formidable ocean swell. Mountains of seawater, rolling toward the broken shoreline, would break up the ice-field, causing gigantic icebergs to dance about in the water like cork buoys. The hollow detonations of these colliding giants sounded like thunder. And the little schooner, its sails furled, would find itself in flight, tossed about like a toy in the prodigious aftermath of a seismic shock.

Throats tightened and the men exchanged worried glances in the half-light of the forecastle. It was usually only a matter of time before the hull struck. A sudden pitiless tearing sound would be heard throughout the vessel as the men found themselves being hurled from their bunks beneath the violence of the shock. At the mercy of the waves, the helpless boat would then tear itself apart on the ice.

At such moments, it was every man for himself. The entire crew would climb into the rigging and cling to the yards, while awaiting the right moment to leap onto one of the icebergs that filed past them like huge buildings, swiping at the wreck as they passed.

Newfoundland and Region

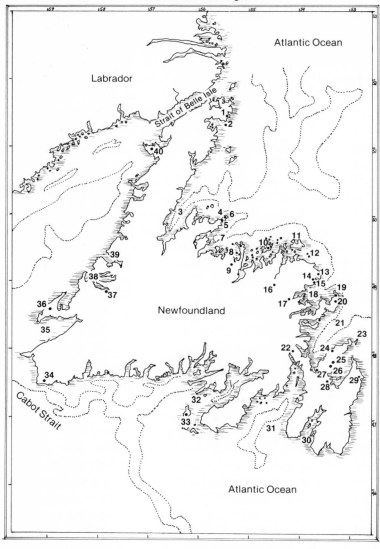

1 — St. Anthony	15 — Greenspond	29 — St. John's
2 — Hare Bay	16 — Gander	30 — Saint Mary's Bay
3 — White Bay	17 — Gambo	31 — Placentia Bay
4 — Canaille Cove	18 — Bonavista Bay	32 — Fortune Bay
5 — La Scie	19 — Bonavista	33 — Saint-Pierre and Miquelon
6 — Cape St. John	20 — Catalina	34 — Port-aux-Basques
7 — Notre Dame Bay	21 — Trinity Bay	35 — Saint George's Bay
8 — Bay of Exploits	22 — Come By Chance	36 — Port-au-Port Peninsula
9 — Lewisporte	23 — Baccalieu Island	37 — Corner Brook
10 — Twillingate	24 — Heart's Content	38 — Bay of Islands
11 — Fogo Island	25 — Carbonear	39 — Bonne Bay
12 — Wadham Islands	26 — Harbour Grace	40 — Port-au-Choix
13 — Cape Freels	27 — Dildo	
14 — Wesleyville	28 — Brigus	

Sometimes, a man would jump too soon or from too great a height and would break his legs or his neck; sometimes, there was nothing for him to cling to but the sheer wall of ice and he would tear his nails out and bloody his fingers trying to get a grip, only to find himself being hurled from the giant. It isn't hard to imagine the countless hideous ways a man might meet his death out there on the ice! At times, petrified with fear, a man wouldn't jump at all and would go down with the boat, teeth clenched, arms wrapped tightly about the mast-head, trembling at the horrible spectacle of the cold, white wastes of eternity that stretched before him on the other side.

But there is a god who watches over sealers, just as there is a god who watches over drunkards. Tragedies were astonishingly few, despite the ever increasing numbers of schooners competing annually in Newfoundland's great national sport. This was a miracle, for most of the sealing captains were rather inept seafarers outside the ice-fields. Their knowledge of navigation was limited to a few rudimentary principles. They knew they had to take a north-easterly course to reach the ice; they knew that the latter was moving southward and that, consequently, they had to steer in a north-westerly direction to make home port some weeks later; and that was about all they knew. And their pride always prevented them from sailing in the wake of another ship!

The tales of lost captains have all the flavour of a "Newfie" joke, though they have the advantage of being true. One such captain, believing he had caught sight of Ile Baccalieu in the fog, hailed a passing boat from Miquelon to inquire if this was true; when informed that, no, it was Saint-Pierre, which lies some two hundred miles south-west of Baccalieu, he actually thought he had reached France! Another, Caleb Sands, became a legend in his own lifetime for his notorious lack of sense of direction; only those men who couldn't find a berth on another boat would sail with him. One spring, luffing under full canvas for the "northern ice," he found himself unexpectedly in Sandwich Bay, high on the Labrador coast. His disheartened crew abandoned him there, making their way homeward on foot.

Of course, boats *were* lost to the ice. Records set their number at about three hundred for the first half of the nineteenth century – about a half-dozen per year, or one per cent of the sealing fleet. And, inevitably, men went down with them. But as long as the sealers remained their own masters, the toll in human life did not seem excessive – no more than the number of fishermen swallowed each year by the voracious sea.

As long as the sealers owned their own vessels, which, as they were also used for fishing, were their only means of making a living, they avoided taking unnecessary risks. Moreover, the boats were small in those days and the loss of a craft with all hands rarely resulted in the death of more than twenty men. Wrecks were caused largely by easterly winds, when the thick ice of the polar cap was blown toward the shore; if the shipwrecked men were patient, they could generally make their way back to land without even wetting their feet. Alas, however, for those who tried to regain the shore as quickly as possible, rather than wait for the tide to carry them in safe and sound on their slab of ice! These were invariably the ones who were lost.

Meanwhile, the seal catch continued to grow. The supply seemed inexhaustible, and the necessity of leaving more seals on the ice than could be taken back on the small schooners was a source of great frustration to the sealers. Around 1830, their appetites whetted by a quarter-century of sealing, those captains who had been able to accumulate a few savings began to equip larger, more solid boats for the ice.

At first, they were ridiculed. Those big vessels would never manoeuvre successfully through the narrow leads! They would never climb onto the ice! Everyone predicted the worst. But when they returned triumphantly the first year, loaded to the hatches, their critics were forced to eat their words, and the fashion shifted gradually to larger vessels.

This was to alter not only the appearance, but also the very nature of the sealing campaign. The new brigs, capable of transporting cargoes of up to a hundred tons, did not lie within the means of every man's purse; not even the most prosperous captains could finance them alone. So they were obliged to go into partnership with the merchants; and since the latter were not known for their honesty and were quick to take advantage of a good deal when they saw one, they were not long in gaining the upper hand.

Manned by a crew of forty men, the new brigs could bring back fabulous catches of up to ten thousand pelts each year!

The *catch*! What sealer, fisherman, hunter, buccaneer, pirate, is not inflamed by the very sound of that word? It is the supreme reward for all the efforts, sufferings, and discomforts endured in earning one's living from the sea. Next to it, even those big pay cheques at the end of an unusually successful fishing trip pale into insignificance (as witness the urge of so many fishermen, upon setting foot on shore, to squander their earnings in drinking and carousing!). But the *catch*! No amount of drinking or revelry can erase the intoxicating memories of the net full to the brim, floods of cod spilling onto the bridge, the euphoric sensation of sailing a vessel full to the hatches, sitting low in the water, riding the waves as if it were part of the great ocean swell itself.

The owners of the new brigs had no trouble recruiting crews. Blinded by the size of the anticipated catch, the sealers seemed joyfully oblivious of the fact that it was no longer theirs and that they were quickly becoming the instruments of a new dynastic class that would eventually exercise its formidable power all along the eastern shores of Newfoundland.

They avidly vied with each other to man these big vessels. There was such a surplus of manpower that the owners profited by obliging the men to "rent" their berths in advance – if you can call the rough wooden planks that served as mattresses in their unheated quarters, berths! Once he had handed over his berth money, the sealer became the proud possessor of a ticket entitling him to passage out to the ice; and these tickets became the object of a flourishing trade on the black market: they were resold for astronomical sums, they were exchanged for better tickets on brigs commanded by better captains, they were stolen, they were seized by force, they served as barter in procuring political favours! So many of those who couldn't obtain a ticket embarked as stowaways, that it came to be considered a bad omen not to have at least one on board every ship. Once discovered at sea, stowaways were made to slave even harder than the regular crew, without pay; but they seemed to prefer this to not particpating in the seal hunt at all!

The feverish anticipation of the catch exercised such a magnetism on the sealers that it took them nearly a quarter-century to realize that the rental of berths was tantamount to putting down deposits on their own slavery!

For what did these famous tickets entitle them to?

A month before their departure, they had to begin making the brigs ship-shape for the ice: covering the prow with a sheath of copper or greenheart wood, shoring up the timbers, mending the sails and the rigging, tending to the thousand and one minor repairs that always precede a sealing campaign. Not only were they not paid a cent for all their pains, they were not even fed!

At sea, they had to work as sailors, perched high in the rigging, in the icy polar winds, unfurling and clewing up the sails. They had to rally for hours on end to extricate themselves from slob ice, cut ice, hitch themselves to hawsers and drag like horses. They had to toil and sweat and drudge till they were ready to drop. In return, they were served meagre rations of raw potatoes and hard-tack, sea biscuits prepared from a mixture of flour, lard, and water that were invariably as hard as brick; for it was the merchants' belief that men with full stomachs were inclined to be drowsy and lazy.

On the better boats, a hot meal of potatoes and salt fish was served once a week. Of course, the captains and officers enjoyed hot meals as a matter of course.

When the famished men reached the seals, they indulged in orgiastic feasts of blood and raw meat, drinking the blood straight from the gushing arteries and tearing with their teeth into the warm, still-beating hearts of the whitecoats. The hearts were strung onto cod lines and worn about their waists, as trophies . . . or snacks. And then they went to work, striking, sculping, dragging the pelts back to the boat, stowing them in the hold, slaving from sunrise till long after sundown for days on end without rest, the same endless, exhausting toil.

When the hold was full, the crew's quarters were requisitioned to carry additional cargo. The men had no choice then but to sleep on the frozen deck, beneath the stars, up to their ears in rancid blubber, even though each man had rented his berth!

Upon their return, they had to unload the catch, scrape the fat from the sculps, melt down the oil, and pour it into barrels. Only then did they receive their wages, which, after all the deductions never amounted to much. Over and above the rental of their berths, they were obliged to pay for their hunting gear – knives, whetstones, ropes, gaffs – the inedible food they had been served, and even the barrels in which the fat had been stowed!

However, misery was rampant in all the little outposts from which the sealing fleet set sail: St. John's, Brigus, Harbour Grace, Carbonear, Wadham, Fogo, Twillingate, Wesleyville, Greenspond, all the way to Battle Harbour, Labrador. Death from illness (beriberi, tuberculosis) or by drowning was each man's constant companion. Even today, they say in those towns that the sea is made up of the tears of grieving mothers. Slavery disguised as a sport seemed a mere inconvenience alongside such ills.

It wasn't long before the carnival of blood reached full force, as industrial methods supplanted the old artisanal ways. By 1830, the great nation of pagophiles was being depleted annually by more than a half-million head: 558,948 in that year; 686,836 the following year; 508,407 in 1832; 631,375 in 1840; 685,530 in 1844; reaching a peak in 1857, when 13,000 Newfoundlanders, almost the entire able-bodied male population of the island, brought back more than 700,000 pelts.

And they were not the only ones to exploit the bounty. Alerted to the formidable resources of the Front, the Scottish whalers of Dundee and Peterhead began to arrive each year to try their hand at sealing, before taking off in pursuit of the cetaceans in Baffin Bay. Of course, they were not as successful as the sealers from the eastern shores, being ignorant of the art of copying – crossing a lead by jumping from ice pan to ice pan, much like mountain goat jumping from rock to rock, which a man must have practised from early childhood to have mastered. Nevertheless, the Scots made considerable dents in the herd: 48,000 pelts in 1847; 74,000 ten years later; 115,000 in 1881, 152,000 in 1884.

Needless to say, as the catch reached such formidable proportions, the loss in human life also rose to unprecedented levels.

In his book on the seal hunt published in 1927, Levi G. Chafe, the greatest authority of his time on all matters pertaining to the seals, provides a list that is far from exhaustive of the disasters that, like festering sores, stud the history of the sealing campaign. A cursory examination of these statistics makes it immediately apparent that the number of catastrophes drastically increased when control of the hunt passed from the sealers to the merchants. While the author notes the disappearance of only a single boat before 1840 – the schooner *Active*, in 1823, with 25 men on board – there is scarcely a year after this date in which a major disaster of some sort did not occur. In 1840, the brig *Catherine* struck an iceberg and sank, taking 38 men to their deaths. In 1848, the brig *Hibernian* ran aground; 20 men, half the crew, drowned trying to reach shore. In 1852, the brig *Hammer* foundered with 37 men on board. The list goes on and on.

For a time, the inhabitants of the little outports remained

passively resigned to these calamities, allowing themselves to be enslaved body and soul to the merchants' tyrannical regime; but, eventually, grumbles began to be heard about the barbaric treatment of the sealers on the large boats and the small value attached to their lives in the great swiling game. In 1860, some three thousand sealers went on strike at Brigus, one of the ports from which the Newfoundland sealing fleet set sail. Their leader was Captain Dick Supple and their demands were modest: they simply wanted an end put to the infamous rental of berths, which bled the men of one-tenth of their prospective earnings even before they set foot on the ice – and for which their families were never reimbursed in the event that they failed to return!

Everyone agreed that it was abominable for a man to have to put up an ante when his own life was at stake; and besides the courageous Dick Supple, the fifty or so captains preparing to leave for the ice tacitly supported the position of the men. All but one: William Whelan, captain of the *Hound*, a man as legendary for his intemperance as for his iron fist, who succeeded in persuading his crew to part company with the main body of strikers and to prepare to set sail.

But the strikers controlled access to the quays and would not allow anyone to pass. There were scuffles and blows, at first with fists, then with clubs, picks, and axes; until, finally, there was the sound of gunfire. Under the cover of protective fire, the scabs succeeded in boarding their vessel, and as the enraged strikers were preparing to take the *Hound* by force, William Whelan ordered the moorings to be cut. The boat drew gently away from the quay, carried toward the middle of the channel by the ebbing tide.

The strike had been broken. Already, the crew of the *Hound* was making sail: the sheets fell from the yards and the spanker climbed the mizzenmast, while the jibs billowed in the wind and the brig quickly began to make headway.

This departure, in the wake of a violent confrontation in which a number of sealers had met their deaths at the hands of other sealers, broke the solidarity of the strikers. They continued to hold the quays for a while, but without conviction. The captains began to show signs of impatience: if they delayed their departure for the ice any longer, the whitecoats would have finished moulting by the time they reached them. Then they would have to be hunted in the water with rifles, a costly procedure that never resulted in a large catch.

But the men were reluctant to leave without having gained some small concession from the merchants. A week after the outbreak of violence, the latter agreed to cut the rental of berths by half, which was only reasonable, since the duration of the hunt had been severely curtailed. The captains set sail, little knowing that not one of them would return with a single pelt.

As for the *Hound*, it sailed back into port a few weeks later with a hold full of blubber.

XI
Steam and the Decline

The following spring, 1861, the policy of renting berths was reinstituted without a word of protest from the sealers. The loss of an entire spring on the ice, of a season's earnings – however paltry and out of keeping with the suffering and work invested – was a rude blow to the precarious economy of most Newfoundland families, for whom poverty was a way of life; and no one could afford the luxury of coming to grips with *das Kapital* two years in a row. To do so would have been suicide.

Since they had first taken an active interest in the swiling game a generation earlier, the merchants of Newfoundland had reigned supreme. One by one, the last of the free agents and private boat-owners were eliminated from the race, and the hunt fell under the exclusive control of a coterie of affluent families. Only in the north, at the extreme tip of the Great Northern Peninsula, did it retain its old artisanal and communal flavour, which has survived until the present day. Elsewhere, the merchants called all the shots in this increasingly flourishing business. The oil market was booming, and seal by-products were being exported by the wealthy Newfoundland oligarchy to more than fifteen countries, including China.

Meanwhile, the size of the boats continued to grow, as their numbers diminished. Eventually all sealing activity came to be concentrated in four large southern ports: Brigus, Harbour Grace, Carbonear, and St. John's. Later, following the conversion of the fleet to steam power, all activities related to the "prosecution of the seal hunt," as the Newfoundlanders put it, would be centred in St. John's, the nation's capital.

However, the hunters themselves continued to hail from the outports, where natural selection played such an implacable role in the lives of the inhabitants that it was necessary to be made of steel simply to survive to adolescence.

Each year, the seal hunt gave rise to formidable migrations of manpower toward the southern ports. Between mid-January and mid-February, the men would set out from the little outports nestled in the depths of the northern fjords, wretched little settlements made up of a few ramshackle hovels, stages and a scattering of fish flakes clinging to the rocks, all permeated with a perpetual odour of brine and fish. At dawn, they could be seen descending the rickety ladders and staircases that led down to the flimsy stages; then they would push off, rowing hard for the mainland. There they took to the road that wound around the deep bays and that brought them, several days later, to the ports from which the sealing fleet set sail.

From as far away as La Scie and Canaille Cove, from Exploit Bay, Gambo, Dildo, and Clarenville, Bonavista, Catalina, Heart's Content and Come by Chance, long columns of sealers followed in each other's tracks along the snow-covered trail – a procession of dark, emaciated silhouettes, their few belongings tied up in a bundle on the end of a stick; sometimes, a sack of potatoes or turnips thrown over one shoulder. They flowed in from everywhere, bundled up in sealskins, their cheeks hollow and their eyes glittering, the flaps of their caps pulled down over their ears. There were adolescents barely out of childhood, mature men, old-timers with snow-white hair; and, one by one, their silent troops merged to form a great legion moving ever southward. For hundreds of miles, the trail bore the marks of their passage: nests hollowed out of the snow where they had slept under the stars, congealed streams of brown tobacco juice, and footprints, millions of footprints.

The arrival of this army of vagabonds, thousands of hungry men without money or lodgings, sowed panic in the villages through which they passed and in the bustling little market towns that had sprung up around the harbours of Conception Bay, where the men put up for a few days before leaving for the ice. Forgotten was the fact that this was the "heroic army of the ice," the very pride of Newfoundland, the harbingers of certain prosperity for the southern part of the island. When the hordes of starving men were spied in the distance, children were hurried indoors, doors were bolted, the residents played dead. Everywhere they went, the sealers were treated like the plague.

Such was the gratitude of the privileged for those upon whose very blood they fed!

Only in the meanest haunts, where the bonds of kinship were woven by the fleet hands of physical and spiritual depravity in an atmosphere of the most sordid misery, did they receive a welcome; and, even there, it was only for the few pennies still lurking in the bottom of their pockets. In return, they were offered a shot of contraband rum and the body of a sickly lass with a toothless grin, then quickly sent on their way.

In the cold, damp quarters of the boats, the evenings had a more jubilant character. After a long hard day's work, the sealers step-danced to the chords of an accordion and sang the interminable laments of the sea. Then, as they began to grow drowsy, the old-timers would tell the harrowing tales of the swiling game, stories to make your hair stand on end. Curled up on their rough wooden benches, the neophytes would listen breathlessly, stricken with terror.

A favourite was the epic of the men of the Wadham Islands, a classic in the oral literature of Newfoundland, dating from the era of the open boats a century before.

One spring, three small crafts from Wadham were carried off in one of those frightful storms that are often unleashed without a moment's warning at the approach of the equinox. The sea was thick with slob ice and whipped by hail and hurricane-force winds, and the men were convinced they were lost until, suddenly, above the whistling and roar of the cyclone, they heard the even more deafening sound of waves breaking against rocks. A spark of hope was ignited in them then, for they thought this must be Funk Island, a high, sheer, sinister rock, the refuge of myriads of sea birds, that sits out in the Atlantic some thirty miles from their archipelago – the sentinel of the Newfoundland shores. They made for it, knowing it was their sole chance of survival. If they missed it, they would be blown straight out into the Atlantic.

But the seas were fearfully rough. The gigantic ocean swell rose right to the top of the islet, where it broke in great sheaves of spray, and rushed in foaming torrents into all the hollows and crevices of the rock, filling the air with a thunderous din. To attempt to land under such conditions was suicidal, but they had no choice. They risked it, and lost two of their three boats in the process. For three days and nights they waited for the storm to blow itself out, without food or shelter, clinging to the bare cliff, while the icy seas slashed at them like knives, chilled and mute in the face of the terrible majesty of the enraged elements.

When the storm abated, they immediately put out to sea again, twelve men in a small boat designed to carry three, with their daily catch of cod! But they had misread the weather, for the calm was only the eye of the cyclone, and they had gone no more than a few miles when the storm raged down on them again in all its fury. All about them the waves rose like giant pyramids, then sank into abysmal chasms, and the little boat danced along, from peak to peak, threatening at any moment to founder, so low did it sit in the water. Then it began to snow, a thin, sharp granular snow characteristic of glacial nor'westers that can obliterate everything at a moment's notice.

For two more days, the men from Wadham bailed and struggled to keep the craft afloat. Then, on the third night, they looked up and saw above them through the torrents of snow, a huge white bird, the like of which they had never beheld. It was a gigantic creature, like a cross between a solan goose and an eagle, though much larger than either, with enormous opaline eyes and talons that looked as if they could seize the boat and its twelve occupants and fly off with them into the storm. All night, the monstrous creature wheeled about the boat, emitting shrill death cries. Then, at dawn, it disappeared into the west.

At that very moment, the storm abated.

For these men, who were naturally superstitious – and who were prey now, too, to the lightheadedness of hunger – this apparition could only be an omen; and as they were far out to sea, with no navigational instrument to direct them, they decided to steer in the direction in which the bird had fled. Their decision proved an auspicious one, for the same day they reached the Wadham Islands.

Since then, as a token of their gratitude, the men of Wadham have traditionally refrained from hunting white sea birds.

Supernatural visions, premonitions, the intervention of occult powers: these were common occurrences on the ice when disaster was about to strike; and each evening, new stories echoed those of the night before. There was the young man who dreamed one night of being carried out of his home in his coffin and who, the following morning, refused to set foot on the ice. Not one of his crewmates returned to the boat that evening. There were tales of monsters: a giant seal, larger than a cystophore, the king of pagophiles, rising up before the sealers and commanding them in the voice of their captain to return at once to the boat; a great horned beast, resembling a bull or a giant musk-ox, appearing to the men to warn them of imminent danger. Some had seen the ghosts of long-lost sealers rising out of the ice, their faces swollen and blue, ribbons of flesh floating from their gaunt skeletons, putting them on guard against some unforeseen catastrophe. It made little difference what form the mirage took, the moral of the tale was always the same: those who heeded the omen were saved, those who didn't went straight to their deaths!

And so the evenings passed in the cold, damp quarters of the boats, while the men awaited the moment of departure for the ice. The purpose was to instil fear in the neophytes, as if in that way to acquaint them once and for all with the frightening reality of the sinister industry which the seal hunt had become. It was said, in those days, that a man must look death straight in the eye at least twice to merit the name "sealer."

Death: it seemed to be the very essence of the swiling game. In the minds of the merchants and shipowners, the death of the seals and the death of the men who hunted them were intimately mingled. To them, hunters and hunted alike belonged to the same exploitable species, and they considered their immolation on the ice-fields a sacrifice necessary to the welfare of the industry and, consequently, to the prosperity of Newfoundland itself. Through an aberration of reason common to many privileged people, it never entered their minds that the sealers were *also* Newfoundland.

And while the men, whom they didn't even have the decency to feed, were busy outfitting their boats for the trip out to the ice, driving innumerable nails into the greenheart planks with their heavy mallets, the owners strutted about on the quays like restless beasts, skinflints puffed up with a sense of their own importance.

Holding a monopoly on the business, they also exercised a monopoly on the sport, and as they were not physically involved, they practised it with all the competitive ardour at their command. It was a harsh, pitiless, vicious business, though watching them stand about in little groups on the wharf on a fine morning in late

February, bundled up to the ears in expensive furs and chatting amiably, you would have sworn they were the best of friends – and perfect gentlemen to boot. Their words were drowned in the hubbub of the busy harbour, but they probably differed little from the idle chit-chat of race-horse owners prior to a big race: the merits of the horses, the price of oats, the condition of the jockeys, the odds. Meanwhile, they gave the captains strict orders never to go to the aid of a rival in distress – unless, of course, it was to make off with his catch and crew.

By an unwritten agreement, the sailing date was set for March 1. Later, when the steam-powered vessels made their appearance, and later still, with the introduction of steel-hulled boats, there was a gentleman's agreement that the steamers would not leave before March 10 nor the steel-hulled boats before March 14, to give everyone an equal chance. Along with the prohibition against hunting on Sunday, the Lord's Day, these were the only regulations governing the campaign.

At the approach of the magic date, the little harbours from which the sealing fleet set sail became hives of excitement. No pains were spared by the merchants and shipowners to demonstrate to the public that they had the economy of the island at heart. On the last day of February – the birthday of the seals, the Newfoundlanders call it, for it is about then that the whitecoats of the Front are born – the merchants held a great feast in honour of the captains. On this occasion, everyone swallowed his pride and his spirit of competition to present to the assembled notables the spectacle of solidarity in pursuit of the common weal. There was the blessing of the boats, followed by the official reception of civil and religious dignitaries, an ecumenical Te Deum, pompous speeches filled with eulogies for the "valiant army of the ice" and the "glorious soldiers of industry," and finally a great banquet, attended by all the local dignitaries, which lasted late into the night with its endless toasts to the unprecedented success of the hunt.

And while the captains were being flattered and entertained in this atmosphere of flagrant fraternity – for it was *they* who were the heroes of the hour – their crews, the "valiant army of the ice," excluded from the festivities, offered an even more edifying spectacle of solidarity on the eve of the great adventure, as, dripping with sweat, they slaved to saw a channel through the ice-bound harbour to the open water!

The day of departure, with its sour morning-after taste in many mouths, was the occasion for a great turn-out. With the town's dignitaries at their head, the inhabitants of the ports invaded the wharves, cheering, waving handkerchiefs, holding back tears, shouting farewells, joking with the men who stood crowding the decks of the vessels, returning their vociferous greetings with jokes and boasts and salutations of their own.

Though charged with excitement and emotion, the atmosphere did not lack a certain decorum: the boats were draped in flags and bunting and for once looked almost clean; and the townsmen in their Sunday best presided over the ceremonies with a certain awkward stiffness. Bands of bugles and bagpipes played martial airs, and there floated over the proceedings that sense of intoxication that often pervades large gatherings, when for a time all differences are forgotten and everyone seems momentarily at one.

A final speech, the moorings were cast off, and all the pageantry and euphoria died abruptly away to make way for the more prosaic vision of the swiling game. Leaping over the railings, hundreds of men harnessed themselves to hawsers and strained like horses to pull the heavy, stone-ballasted vessels through the narrow channel they had cut the previous day. A hush fell over the crowd as the gap separating it from the boats widened, that sense of emptiness that always opens up between a group of people on board a departing vessel and those back on the quay. When they reached open water, the brigs set sail and slipped one by one toward the distant horizon. Hearts back on shore were heavy. Swallowing back the lumps in their throats, the citizens exchanged worried glances, haunted one and all by the suspicion that they might never see these men again.

The advent of steam in 1863 didn't greatly alter the pattern of the hunt, except to exalt the element of death in which it was already firmly rooted. But, in the end, death did not remain the prerogative of the seals and the sealers alone; it turned against the industry.

The acquisition of steam-powered vessels by the Newfoundland merchants was the result of circumstance rather than of any deliberate strategy on their part. The year 1862 had been a very unprofitable one for the sealing fleet: the winds, blowing incessantly out of the east, had prevented the big sailing vessels from reaching the main patch and they were forced to be content with small peripheral catches. This in itself would not have been cause for undue alarm if another event hadn't occurred to further incense the frustration of the captains: while they waited, powerless, for the wind to change, they could see the fleet of Scottish whalers in the distance, manoeuvring with utmost ease through the ice-field. Having approached the ice from the east, it was easy enough for them to navigate under sail, but, since two of their vessels were equipped with steam engines and spewed thick torrents of black smoke into the air, it was to this unprecedented phenomenon that the Newfoundland captains attributed their success – wrongly so, as it turned out, for neither of these steamers took a single seal that year. The sailing vessels fared little better, despite their mobility, for the winds had piled the herd up against the shore. It was strictly a landsmen's spring that year.

Nonetheless, the myth of the fabulous catches made possible by the advent of steam captured the public imagination. That summer two of St. John's bigger merchants – Walter Grieve and Baine, Johnson & Co. – sent their agents to Scotland to look into the possibility of acquiring these miraculous boats.

They returned with two archaic vessels, the *Wolf* and the *Bloodhound*, decrepit old whalers, saturated with salt water and whale oil from their long careers in the boreal seas, but equipped with James Watt's precious invention, the steam-engine. These big caravels weren't much to look at, and their antiquated boilers, laced with cracks and spouting steam at every joint, generated little more than a symbolic thirty horsepower, so they often moved faster under sail than steam. The motor of one of these vessels generated so little power that its screw stopped turning when its whistle was blown. But they were the latest thing in progress and modernization in the eyes of the Newfoundlanders and, consequently, became an inexhaustible source of pride for their new owners.

In the spring of 1863, no one was impressed with their performance – they returned from the ice with a combined cargo of no more than four thousand pelts – and it is probable that the conversion of the Newfoundland fleet to steam would have been delayed yet another few years if two new circumstances hadn't intervened in its favour. By this time the whale hunt in Baffin Bay was virtually a thing of the past, the prey having been all but exhausted, and the Scottish shipowners of Peterhead and Dundee were selling off their old whaling vessels at bargain prices. Also, the great herd of pagophiles, decimated by more than twenty million heads since the beginning of the century, was showing signs of depletion, so that it was necessary each year to push farther and farther into the ice-field to reach the main patch, a difficult and perilous task for the big sailing vessels.

Between 1863 and 1872, the merchants of Newfoundland acquired twenty-six old steam-powered whalers, and thus the great hunts of the eastern shores entered a new era, one which would lead them at an ever-accelerated pace to their inglorious end.

These whalers – or "wooden walls," as they came to be known – were built to withstand just about anything. Their seventy-metre keels were constructed entirely of oak and birch, and their planking was sixty centimetres thick. Their prows, covered with a sheath of greenheart wood and an armature of metal plates, were braced with beams nearly a metre in diameter. The bolts that held the armature in place passed through more than three metres of solid wood.

For twenty, thirty years, sometimes longer, they had been driven relentlessly through the ice-choked waters of Baffin Bay, ramming ice-packs under full canvas, subjected to the worst possible treatment as they followed in the wake of the great whales. And for twenty, thirty years, sometimes longer, they then worked the ice of the Front, treated with even less regard by the Newfoundland sealers than by the Scottish whalers. Certain vessels in the fleet, like the SS *Terra Nova*, built in 1885 – the ship chosen by the explorer Robert Scott for his tragic expedition to the South Pole in 1910 – still ploughed the ice-fields half a century and more after their maiden voyages.

In these "wooden walls," the comfort of the crew was sacrificed to the solidity of the hull: all space below deck was taken up by wood. The forty whalers they were designed to carry were lodged in a miniscule binnacle, where half that number might have lived in reasonable comfort. The Newfoundland shipowners went the parsimonious Scots one better: they packed up to two hundred men into this tiny cocoon.

If the living and working conditions on the big sailing vessels had been bad, they were abominable on these new galleys. Since they could carry much larger cargoes, the owners reduced the crew's share from one-half to one-third of the value of the catch. Nothing justified this measure – except perhaps their unquenchable thirst for profits – for all the expenses of the expedition, including the coal burned by the vessel, were deducted from the men's meagre earnings. But the lure of big catches, the magical pull of "progress," had the sealers once again vying with each other for the privilege of working the steamers – forgetting that 45 men sharing 50 per cent of the average brig's catch of 5,000 pelts, made a great deal more than 200 men sharing 30 per cent of the average whaler's catch of 20,000 pelts.

When the Newfoundlanders boast that the era of the "wooden walls" was the golden age of the seal hunt, they are clearly mistaken: it was already the beginning of its swan song.

There was no golden age, except for the merchants – those who survived the savage competition to sweep the ice-floes clean of seals: Baine, Johnson & Co.; Ridley & Sons; Bowring Brothers; A. J. Harvey & Co.; Job Brothers & Co. By 1875, supplies of marine oil were no longer able to meet world demand. The great boreal and austral hunts for the Leviathan had exterminated virtually all the commercial species of whales; the walrus had disappeared from its southern habitat nearly a century before; and the seals of Antarctica, decimated by more than fifteen million head during some fifty years of hunting, were no longer worth the expense of outfitting large sealing expeditions. The only relatively stable source of oil that the London market could rely upon was the harp seal of the North Atlantic. There were fortunes to be made each spring by the merchants of St. John's – the sharks of Water Street, as they came to be called, for their businesses were all located on that street. The frustration they felt at the inexorable diminution of the seal population, they directed against the sealers.

No further precautions were taken to ensure the safety of the men. Since the "wooden walls" were used exclusively now for the seal hunt, they sat idle and neglected at their moorings the remainder of the year. They were never overhauled. Their furnaces were so old and decrepit they could be staved in with a kick. Saturated with oil and filled with highly inflammable seal fat, they were floating firetraps. The slightest spark could send them blazing up like torches. While such catastrophes occurred on three or four occasions, the miracle was that they didn't happen more often.

Moreover, these boats all carried large quantities of explosives to extricate themselves from the ice whenever they became temporarily imprisoned. The crates were stored in a small bin adjacent to the officers' quarters, separated from a stove burning at white heat only by a thin partition. Sometimes, they were kept in the lavatories, where a man, relieving himself, might negligently shake the ashes of his pipe over them. Even after the *Viking* went up like a bomb in 1931, killing 28 men – the worst, but not the only such tragedy – no one thought of taking steps to regulate the storage of dynamite on sealing vessels.

But where the hunt took a truly criminal turn was in the risks the men were made to run on the ice. The "wooden walls" marked the advent of a new sealing technique: the panning, or stacking, of pelts. Each morning, the men were dispatched to the ice in watches of twenty-five or thirty, spread out over a large area of the floes, often far from each other. There, they would break up into smaller teams and begin systematically to work the seals scattered amongst the hummocks. As they finished with one patch, they would pan the pelts, marking the pile with their boat's flag, and move on in search of more seals. Meanwhile, the boat would shuttle all day from one spot to another, retrieving the pans.

In good weather, there were no problems, save the piracy of pans and the occasional loss of pelts that drifted out to sea and were never recovered. But when storms threatened and the captain saw with horror that he risked losing his precious booty in the break-up of the ice, it was a different story.

At such moments, the skin of a man was worth less than that of a seal. The sealers were abandoned to their fate on the ice. Several days later, they might be found frozen to death or dying. Who cared, as long as the sculps were safely stowed below? The merchants made no bones about it. They shipped the corpses that were retrieved from the ice back to their distant outports – at the expense of the bereaved families.

Thus, a diminishing number of boats, the property of a diminishing number of shipowners and manned by a diminishing number of sealers, set out each year in the last third of the nineteenth century to assault the great herd of harp seals, which was also rapidly diminishing.

An epoch, an entire way of life, was passing from the island and no one seemed to be aware of it. Despite the introduction of steam, the catch continued to decline each year. Those great campaigns that had once resulted in the slaughter of more than half a million whitecoats were now a thing of the past. A catch of 400,000 pelts came to be looked upon as an unusually successful one; then it dropped to 300,000.

The year 1884 was an exception: 730,000 pelts were unloaded at St. John's that spring. It seemed for a time that the good old days had returned. But that was an unusual spring, in more than one respect: the whelping ice hadn't broken up until very late in the season and the thirty-odd steam-powered vessels, manned by more than six thousand sealers, had surrounded the entire herd; not a single whitecoat escaped the massacre. In a few days, an entire generation of harp seals was annihilated.

Needless to say, the effects of this holocaust were felt in succeeding springs. It wasn't long before the average annual catch had dropped to less than 300,000 pelts; and, by the beginning of the twentieth century, despite the introduction of steel-hulled vessels with infinitely more powerful engines, it was down to less than 250,000. It continued to plunge to about 125,000 in the years preceding the Great Depression, and to 100,000 at the outbreak of the Second World War.

Meanwhile, how many sealers had also fallen victim to the mad thrust of industry?

XII
Other Times, Other Seals

The departure of Richard Gridley from the Magdalen Islands, followed closely by the extinction of the sea-cows, had left the little Acadian population in a state of mixed euphoria and confusion in the closing years of the eighteenth century. They had rejoiced at the Bostonian's departure from the archipelago, but they were uncertain of what the future might hold in store without the sea-cows. To a large extent, it was the uncertainty that triumphed. This is the paradox of the working classes: with no one to exploit them, masters of their own resources, free to forge their own destiny, they are often inclined to feel helpless. Family by family, the Acadians left the Islands in search of greener pastures, travelling to Gaspé, Quebec's North Shore, Ile Madame, the west coast of Newfoundland, even as far away as Labrador, in a gigantic wave of emigration. Once nearly two hundred strong, the number of Acadian families on the Magdalen Islands had dropped to sixty-eight when the new governor, Admiral Isaac Coffin, took the census there about the year 1806.

The archipelago of the Gulf was the Admiral's little whim. He had caught sight of it, one fine summer's day, while transporting Lord Dorchester back to England from Quebec – little emerald isles strung on a chain of golden dunes, resting like a regal necklace upon the ultramarine velvet of the waves – and he formally requested His Britannic Majesty to turn them over to him in return for all his loyal services. His request was instantaneously granted. To the embarrassment of all, however, it was discovered shortly thereafter that the Islands had already been granted some years earlier by George III to Richard Gridley.

The Admiral manifested great disappointment at this setback, for Gridley was only the colonel of a little artillery unit, while he, Coffin, was a flesh-and-blood hero of the British Navy. So from the moment the Bostonian gave signs of renouncing his rights to the fief by abandoning it, he took steps to acquire it for himself. The transfer of deeds took several years, and it wasn't until 1806 that Isaac Coffin was able to make his first proprietary tour of the Islands. To celebrate the occasion, he had a commemorative penny struck, bearing effigies of the two local resources he intended to exploit: on one side, a cod: on the other, a seal.

His projects did not meet with much enthusiasm from the Islanders, however. Those Magdaleners who had remained on the Islands were strong-minded individuals, and during the few years they had been without a governor, they had acquired a taste for freedom, evolving an idyllic existence based on the classical triad of Acadian self-sufficiency: the sea, the land, and the forest. Following the cycle of the seasons, they were content to work just enough to satisfy their needs, hunting the seal in the spring, fishing and working the land in the summer, raising cattle for meat and sheep for wool, poaching openly in the crown forests, salvaging the wrecks that were occasionally washed up on their shores, exchanging their surplus produce for flour and molasses through the agency of a Jerseyman on Ile Madame – and generally lacking for nothing.

The prospect of reverting to a life of servitude and industrial production did not please them, and they were not hesitant to make their displeasure known to the new governor, who threatened the entire male population of the archipelago with imprisonment if they did not accede to his plans.

It was the fishing of cod, in particular, that the Islanders resented. The rich cod beds of the Islands had long been the monopoly of the Americans, who, with their flotilla of six or seven hundred fishing boats, reigned supreme over the inshore waters. Fearing reprisals, raids against their habitations, and other punitive measures – not least, the end of the thriving smuggling trade that flourished between them – the Magdaleners had no desire to enter into competition with the foreigners. Isaac Coffin settled this question in a draconian fashion by having large fishing schooners built and ordering his subjects to go in search of cod off the Labrador coast, a thousand miles from home!

Axes began to encroach upon the lovely Magdalen forests,

whose praises had been sung by Jacques Cartier, and for nearly a quarter of a century the Islands became a gigantic naval shipyard. Nearly a hundred forty-ton schooners were constructed. By the middle of the century, all the reserves of hardwood had been utterly exhausted. There remained nothing but bare hillocks strewn with alders and stunted spruce, prey to the incessant winds. (Today, "progress" has taken yet another step forward: stripped of their wood, the hills are now being ripped open with bulldozers for their gravel, and the beautiful Magdalen landscape is slowly being blown to the winds.)

Life on the Islands took on a new look under the reign of the Coffins. In the spring, the men hunted the great herd of pagophiles; then they fished for herring, and when they had replenished their stocks of bait, they left on the long voyage for Labrador, their Siberia.

They didn't always return the same year. If the fishing was poor, they would often spend the winter in Cain Land. With the aid of large nets, they would capture the harp seals during their southerly migrations and survive on the meat during the winter. The following spring, they would again find themselves up to their elbows in brine and tripe; and it was not until late autumn that they would wend their way homeward. Meanwhile, despite their absence, they were obliged to pay their annual rent to the governor, under pain of having their lands and homes confiscated.

Iniquitous, inhuman, odious, such was the lot of the early Magdaleners. It was only during the seal hunt that they experienced any real sense of autonomy, the feeling for a time of being their own masters. Out there, on the ice, they were no longer harrassed by the Americans, for the foreign fishing fleets didn't enter the Gulf until after the break-up of the ice; and compared with their long exile in Labrador, their sojourns amongst the seals must have seemed brief and almost pleasant.

It is not difficult to visualize those men of old crossing the ice in single file, zigzagging amongst the hummocks, with their long gaffs and their dogs – black silhouettes against the blinding white backdrop of the living ice like the etchings of Edmond J. Massicotte of a century later – or to see them returning, each man towing a pile of bloody pelts behind him, heaving, straining, while the folks back on shore anxiously awaited their return. Time has succeeded in altering neither the essence nor the complexion of the inshore sealing campaign.

At the beginning of the nineteenth century, this was virtually the only type of sealing practised on the Islands. Most of the men went out to the ice on foot; the *canot* was used only rarely. Near the end of February, a large segment of the population – men, women and children – would pack up their supplies and migrate to the dunes of the north and west, where the great herd generally comes close to land. There, they would erect little makeshift structures and await the seals.

When the herd was within reach, its arrival would be signalled along the dunes by a prearranged series of gunshots, and the men would leave for the ice.

On the north and west sides of the archipelago, where the prevailing currents and winds pack the ice-floes up against the dunes and the ice is solid, its drift virtually negligible, a man can venture several miles out to sea without ever having to cross a lead. This accounts for the name by which this section of the ice-field is known: the "glacier." Seven generations of Magdaleners have hunted the seal there, and I have never heard of a single man losing his life.

The only dangers the sealers face out there are blizzards, blinding sleet storms that come up without warning and cause them to lose their bearings, and sudden shifts in wind that detach the living ice from the shore ice and blow it out to sea. Of course, the seals often remain far out on the floes and the men have to walk long distances to reach them.

Those old-time sealers were well aware of these dangers; they had learned to provide for them. While they were working the seals, the women stayed back on the dunes, keeping an eye on the wind. If it should shift suddenly, they would warn the men with a series of gunshots, a signal to retreat. If the shots were few and far between, the sealers could take their time returning with the day's catch, for the ice would not break up at once. But if the detonations crackled in the distance like fireworks, there was not a moment to lose: they must abandon their pelts on the ice and return on the double.

To find their route in the blizzard, in the days when magnetic compasses were not yet available in army surplus stores, the men relied on dogs. The only reason they took them out on the ice was that a dog will always instinctively head for land even when it can't see where it's going. I learned this one day from an old Pointe-au-Loup sealer, who took his antique grapnels down from a nail in the barn where they had hung rusting for many years, to let me savour the charms of old-time sealing – and, no doubt, to delight in them one last time himself. How his face lit up when he saw that I had brought along my faithful Zoé! This lent the occasion an even closer affinity with his memories of the old days.

While the hunt on foot continued virtually unchanged until the introduction of the skidoo, other types of sealing were evolving on other parts of the Islands. The Acadians of Bassin and Havre-Aubert, as well as the Englishmen of Old Harry and Grosse Isle, got into the habit of paddling little *canots* out to the whelping ice off the south and east shores of the archipelago. After the whitecoats had finished moulting, they continued to patrol the leads in search of bedlamers and adult seals, which they hunted or "swatched" with rifles.

When the fishing schooners began to leave the shipyards, they were also used in the seal hunt. They spent the winters at anchor in the natural harbours of La Pointe, Havre-aux-Maisons, and Havre-aux-Basques, where the currents and tides keep a channel free of ice almost all winter. As soon as the herd had drifted past the Islands and the inshore campaign was over, the sealers would pursue the beasts on these schooners, swatching them as far away as Cabot Strait, sometimes even out into the Atlantic.

No effort was spared to ensure that the harvest should be a good one, for then the men would not have to return to Labrador to fish for cod to make ends meet at the end of the year when they went to settle their accounts with the governor's agent. They not only hunted the seals, they also fished for them – all species of them. Ten years of slavery slaughtering sea-cows had transformed the Magdaleners into a people of the marine pinnipeds!

The net was never much in favour with the Islanders; the men who fished for seals generally preferred the trawl-line – an extremely strong ground line, solidly anchored to resist the efforts of a quarter-ton of muscle fighting for its freedom, from which there extended a number of snoods ending in enormous baited hooks. The trawl line was stretched in the shallows at the mouth of a stream, where the smelt swarm, or along a sandy beach beneath the ice, where the young shrimp abound. Though it is widely believed among scientists that seals do not eat while they are mating, breeding, and moulting, there would seem to have been more than a few exceptions to this rule. In fact, judging by the catch, the amphibians were less conscientious about observing their fast than were the Magdaleners! Hundreds of adult seals were caught on these hooks each spring.

This was not a pleasant fate for the poor beasts. If they had to die, they would surely have preferred being clubbed to expiring slowly and painfully, a few fathoms down, with a hook implanted in their jaw or esophagus wall. Their death struggles must have been long and agonizing, for a harp seal can remain without air for up to twenty minutes.

Trawl-lines were outlawed in the early 1960s, as a result of the public outcry over the slaughter of "baby" seals. Some obscure civil-servant, nauseated by the sight of blood, must have decided they were too cruel. (Maybe he was secretly in love with Brigitte Bardot and didn't want to be mistaken for some barbaric Canadian in her dear eyes?) Whatever the case, the regulation testifies to a certain mental disorder on the part of the individual who drew it up: to forbid the fishing of seals with trawl-lines and, at the same time, to sanction their capture with nets. What, may I ask, is the difference between being strangled in a net and suffocating at the end of a hook? In both cases, the beast succumbs to asphyxiation at the end of an identical period of time!

I feel great sympathy for all the old seal-fishermen (it was generally the men who were too old to go out on the ice who tended the trawl-lines), for they were deprived not merely of an occupation, a pastime, a source of a little income, with the passage of that ridiculous by-law. They also lost an honest trade, an art that may disappear forever for want of being passed on to succeeding generations. And that is regrettable, for it was no easy matter to catch the pagophiles with these lines; not every man who turned his hand to it was successful. There is a special way of attaching the bait to the hook, a special way of laying the line in the interplay of tides and currents so the bait will take on the appearance of a living prey – an art based upon long observation and an intimate knowledge of the habits of the seals, and that calls upon all the patience and ingenuity of the fisherman. If there is the slightest flaw, the crafty pagophiles will scorn the trawl-lines like the plague.

In the absence of statistics, it is difficult to determine the number of seal pelts the early Magdaleners harvested each year. One thing is certain, however: the catch was in proportion to the population. If figures dating from the beginning of our century can be trusted (they indicate that some 3,000 Magdaleners, from 600 familes, harvested an average of 20,000-30,000 pelts each year), then the 68 families still residing in the archipelago at the beginning of the Coffin regime must have slaughtered fewer than 4,000 seals annually; and the 300 families of the succeeding generation, perhaps four times that many – a trifling loss for the great Gulf herd, whose annual birthrate numbered at least a half-million head.

Half of this catch came from the inshore campaign, the other half from the schooners and the trawl-lines.

During the first sunny days of May, the fat was scraped from the pelts and melted down in large open vats equipped with spigots to allow for the drainage of rainwater and snow. Then the oil was poured into kegs and sold to the governor's agent for a few pounds sterling, of which each member of the crew would receive a few shillings. Since all the work was carried out on a communal basis and relied exclusively upon artisanal techniques, this was very small recompense indeed for all the time and effort expended.

Of course, there was always the meat. Returning from the dunes, the men set to work curing it. They cut large blocks of ice from the shore ice and filled their ice cabins with them. Little huts of unhewn timbers located in a shady part of the undergrowth, these cabins served as refrigeration units until the height of the summer heat. There, the carcasses of the seals were suspended to keep them fresh; then, before the meat could spoil, it was salted. In a good year, the sealers could cure enough meat to see them through to the summer.

In the fall, they survived on the meat of common seals, wild geese, and black ducks. They slaughtered an ox or a pig for the great feast days of Christmas and New Year's. Then, at the end of January, following the harvest of the grey seal pups at Deadman Island, they again savoured the distinctive flavour of seal meat. In one form or another, the seal served as a staple all year round.

At the beginning of the Coffin regime, the Magdaleners were much more dependent upon the seal for their survival than they are today. They made boots, tuques, mittens, and coats from the tanned skins; they used the oil in their lamps, and they ate the meat. This heavy dependency upon the pagophile lasted until about the middle of the nineteenth century.

Then, gradually, as a certain prosperity began to appear here and there in the archipelago and social classes began to appear within the ranks of the serfs themselves, the mere idea of dressing in sealskins and eating seal meat came to be looked upon as the lot of the poor, the downtrodden. Rather than passing for paupers in the eyes of the privileged, those who aspired to some social status gradually eliminated the seal from their lives.

It disappeared first from their backs, then from their tables, though they continued to use it in their lamps, lacking a substitute. While the great herd continued to be hunted each spring for the profits it brought to the community, it became fashionable in certain families to tighten belts and to shiver in cloth garments rather than to pass for savages – or, worse, Newfoundlanders! It was then that tuberculosis began to take its toll on the Islands. But what did that matter, as long as appearances were saved?

In this way, the lifestyle of the early Magdaleners eventually became folklore. The people no longer ate flipper stew or whitecoat liver – except during the hunting season, of course, when it was a matter of tradition.

Oh, there were some who continued to dress in sealskins and to eat as much seal meat – and crow meat, too, for that matter – as they could get! But these were, by and large, members of the lower classes of society, a minority generally scorned by the rising middle classes. They lived in shacks on the edge of the hamlets, beyond the immediate influence of the Church, and they were the object of all manner of calumny: they drank, blasphemed, gave themselves up to gambling and vice, sold their wives and daughters to transient sailors, and refused to work, or so the story went! They were seen in the villages only when they came to beg at the doors of respectable homes, where they would draw down curses and maledictions on those who turned them away empty-handed! The sorcerers of Acadia, they were pursued by *Monsieur le curé* with his aspersary of holy water!

The Magdalen Islands are hallowed ground for the seal. Besides the great migrating herds of pagophiles and cystophores that visit the archipelago each spring, two other amphibious species live here on a permanent basis: the common seal or *phoca vitulina*, also known as the bay seal, harbour seal or *loup-marin d'esprit*; and the grey seal or *halichoerus grypus*, which also goes under the name of Atlantic seal, horse-head seal or *rouard*. These two species are the calves and pigs of the sea; and as they differ in nature and appearance as much as their terrestrial counterparts, I shall deal with them separately here.

Common seals are small, timid creatures that live almost exclusively in the water. With their slate-grey backs and silver bellies covered with black spots and haphazardly interlaced white rings, they can sometimes be spotted on beaches and sand bars in the early morning, but the rest of the time they remain in the water, darting after fish and indulging in a great variety of acquatic games. Even at birth, the pups do not remain long on shore: the mothers give birth to them on a sand bar or reef at low tide, and by high tide they have already learned to swim. It is probably for this reason that they lose their foetal fur in the womb or shortly after birth.

These seals are to be found on both sides of the Atlantic, in the temperate coastal waters and the estuaries of large rivers, sometimes even far upstream in the fresh-water rivers and lakes. In 1967, the year of Expo 67 in Montreal, a common seal that had strayed far from its herd was photographed in the waters off Ile Sainte-Hélène near the city of Montreal; apparently it too had come to see what all the excitement was about. Usually, however, common seals are extremely shy, and avoid contact with humans.

The Magdalen colony of harbour seals numbers several hundred beasts, perhaps as many as a thousand. They frequent the inland lagoons and the sand bars and never stray far from home. A mature adult measures a metre and a half in length and weighs up to one hundred kilograms.

The grey seals differ greatly from their cousins in size, stature and temperament. Two and a half metres in length, weighing three hundred kilograms – though the females are appreciably smaller – they strike one as being at once gentle and powerful, with their broad shoulders, their dull grey coats covered with black spots and their snouts that lengthen and widen with age. Save for the ears, the profile of an adult grey seal bears a striking resemblance to that of a young donkey.

In the water, they show no fear of man. One day, while diving for shrimp in the channel of Grande-Entrée, I was approached by two grey seals from a nearby colony who had taken a sudden interest in my presence. They swam about me, now on their sides, now on their backs, now spinning in the water, keeping their distance at first but then coming closer and closer, as if to give the clumsy swimmer I must have appeared in their eyes a demonstration of their acquatic skills. Seeing them dart like torpedoes through the glaucous water, I felt a little uneasy; but I thought it safer to remain where I was than to flee. Suddenly, one of the creatures dashed at me at lightning speed, and I thought my time was up. But just before colliding with me, it gave a strong thrust of its flippers and sped past me, missing me by only a few centimetres! I felt the rush of water against my skin, and I realized that I had got off with a good scare. The animal had merely wanted to play with me, as seals will often do amongst themselves. Being no braver than the next man, however, I decided to return to my dinghy.

On the dark rocks of Deadman Island, during the hunting season, the grey seals are far from being so playful. The big males greet the annual arrival of the sealers with fierce growls and roars, baring their fangs and leaving no doubt whatever of their animosity. More inclined than usual to attack and bite, they can be as fierce as tigers.

It is my belief that this aggressive behaviour is motivated not so much by paternal instincts as by jealousy. During the mating season, which begins more or less with the birth of the whitecoats, the male *rouards* devote their energies to establishing little harems, and bloody combats ensue amongst them over the possession of the females. The sealer who approaches a mother seal to seize her pup, thereby driving her into the water where coitus always takes place, is probably seen by the males as a potential rival.

In the first few years of their lives, the grey seals wander far from their native shores. Taking to heart the proverb that travel shapes character, they depart in search of adventure, exploring the coves and inlets of foreign shores, frequenting the mouths of small torrential tributaries during the season of the great salmon runs, embarking upon long excursions that often last several years. But as soon as they reach sexual maturity, they lose their nomadic instincts and spend the remainder of their lives at home, rarely straying far from the spot where they first saw the light of day.

Long before the arrival of the white man, the Mic-Macs and the other native peoples who inhabited the shores of the Gulf hunted grey seals and common seals.

For them, the seal hunt responded to a two-fold need: first, it allowed them to protect their fishing territories from the devastating assaults of those insatiable fish-eaters; secondly, it provided them with meat and furs. Even today, certain Amerindian tribes in the county of Bellechase, in the Lower St. Lawrence region, live off the sale of articles made from the fur of the common seal.

In the Magdalen Islands, the hunting and fishing of common and grey seals was practised regularly from the time the sea-cows were exterminated until the consumption of seal meat began to fall into disgrace. Thereafter, except for the traditional annual harvest of *rouard* pups at Deadman Island, the animals were left in peace. The armistice was to last until the dark years of the Great Depression.

At that time, sealers from Anse-à-la-Cabane would have made a fortune from the commerce in grey seals had it not been for the quick intelligence of those big pinnipeds, who soon learned to be wary of them. Legend has it that, following several years of heavy losses, the seals learned to anticipate the sealers' moves and to gauge their strength and limits so effectively that they would openly defy them, disappearing beneath the waves the moment rifles were trained in their direction. Others would deliberately offer themselves as perfect targets, with all the accompanying mimicry of helpless martyrs, taking care, meanwhile, to keep just out of range!

After the War, the hunting of indigenous seals once again fell out of fashion. There was no further need to exploit them, and the Magdaleners, more respectful of nature than the happy-go-lucky week-end sportsmen and county-fair marksmen who take such pleasure in squandering wildlife, realized that hunting these amphibians with rifles demanded more skill than most of them would ever possess. Practised by amateur marksmen, it almost invariably results in a shameful waste of life and ammunition. Often, the beasts are shot from too great a distance and are only wounded; or, if they are killed, the corpses cannot always be retrieved before they sink. So much waste and suffering for nothing.

Anthime, Nathaël, Baptiste, Vilbon, Cari, Wellie, Nazaire, Maurice and all the others hung their 303s back on the wall.

It wasn't until 1976 – the year the Canadian government began to offer premiums for the slaughter of common and grey seals: $5 for the jawbone of a common seal, $25 for that of a grey seal (increased to $50 in 1979) – that they took them down again.

This unfortunate measure clearly responds more to political than to ecological imperatives, whatever the Minister of Fisheries and his deputies may say to the contrary; for the common seals and grey seals, whose numbers are now dangerously low, are far less harmful to the fishing "industry" than they would have us believe. The whole business is a prime example of the manner in which the government often caters to the imaginary needs of its electors by offering them things they haven't even requested – a smokescreen to disguise its impotence in areas where its intervention is more desperately needed.

Wellie, who makes it his business to examine the contents of the stomach of every fish he catches and every beast he kills, has never found anything but shrimps, flat fish (rarely fully grown), seaweed, and stones in the stomachs of those seals that now have a price on their heads. This isn't to say that they don't occasionally feed on the commercial species of fish, nor that if a lobster were to lodge itself between their teeth they wouldn't close their jaws on it; but they do not make an exclusive, or even a habitual, diet of them.

And even if they did, the system of premiums would not be any less absurd, for the population of the grey seal colony is already protected from any immoderate demographic explosion through the annual harvest of whitecoats. Certain articles included in the Canadian law for the protection of seals defy all logic. For example, the regulation that shamelessly encourages the slaughter of adult seals with rifles also places severe restrictions on the hunting of pups with clubs. A special permit is now needed to hunt the grey seal pups of Deadman Island. Needless to say, there is no question of premiums being offered for the slaughter of innocent "babies."

Meanwhile, the doors have been opened wide to an invasion of unqualified bounty-hunters, who shoot at anything that moves, blinded by the lure of the premium, and who usually succeed only in wounding the animals. Each summer, dozens of rotting, fly-infested seal corpses are washed up on the Island's beaches.

Let no one misunderstand me: I am not campaigning for the prohibition of seal hunting with rifles. Far from it. If a fisherman feels he has a score to settle with a seal that has stolen his fish or plundered his nets, then let him settle it – that's his affair. And if an Islander has a taste for common seal meat, then let him slaughter one and treat himself to a feast: that's in the nature of things. But the government has no business interfering with vile incentives to indiscriminate slaughter. Its behaviour must seem strange to the rest of the world: this is the same body that spends so much money and energy trying to convince the public that the harp seals are safe from extinction under its vigilant protection!

Officials claim there is another reason for placing premiums on seals. They are bearers of a parasitic worm known as "Terra Nova decipiens" or cod-worm, which gets into the fish fillets, necessitating the employment of extra labourers in fish-processing plants to remove it. But offering rewards for the jaw-bone doesn't solve the problem. Since bounty hunters are only interested in the seal's head, they generally sever it on the railing of the boat, letting the remainder of the worm-riddled body drop back into the water.

Without the existence of these bounties, there is no doubt that the Magdaleners would be much less trigger-happy, for they know better than anyone that, until late in the autumn, the fur of the *rouards* is covered with ugly scars from wounds incurred during the mating season and is worth next to nothing!

XIII
Springs of Mourning

As the nineteenth century unfolded, life continued to have a bitter flavour for the Magdaleners. The sixty-eight families Isaac Coffin originally enumerated multiplied at an astonishing rate, quintupling their numbers with each successive generation, and proportionately increasing the problems placed upon them by the constraints of space and the feudal regime. There was widespread concern, amongst subjects and masters alike, about the drastic decline in the reserves of wood, the bewildering disappearance of game from the crown forests, and the increasingly strained relations between the rulers and the ruled.

In the face of a population explosion, the governor's agents ruled the little colony with an iron hand. They applied, as if they had invented it, the doctrine of divide and conquer; and they stopped at nothing, not even the most odious sort of blackmail, to incite a spirit of envy, mistrust, and suspicion amongst their subjects.

This ignoble system reached its peak during the reign of the agent John Fontana, which lasted from the mid-1830s well into the second half of the century. Of Anglo-Italian stock, Fontana had immigrated to the Islands as a youth and was married to an Acadian girl, a fact which he felt entitled him to treat the Islanders as ignominiously as possible. He made them sign ten-year leases, he raised the rent of their holdings, he encouraged competition and rivalry, he accepted bribes, he rewarded informers with special favours. He worked so hard and so effectively at inspiring a feeling of awe amongst the people that, in the end, his efforts backfired. Instead of remaining submissive, the inhabitants began to grumble and protest; and when he responded to their dissension with an escalation of his hard-line policy, the discouraged Magdaleners embarked upon a mass exodus in search of less tyrannical horizons.

The size of this new wave of emigration can be appreciated by the fact that from 1840 till the early 1870s no less than three hundred families – some historians put the number at closer to five hundred – left the archipelago. The Magdaleners refer to this evacuation as The Small Expulsion or *le Petit Dérangement*; for in its own sad way, it commemorated the centenary of *le Grand Dérangement*.

Like a flock of startled birds at the sound of gunfire, the emigrants scattered along the shores of the Gulf, from the Baie des Chaleurs, Chéticamp and the Gaspé Peninsula, all the way to Newfoundland and Labrador. Far from the migratory routes of the great seal herd, and as often as not immersed in an Anglophone milieu, they gradually lost their sealing traditions, their language, and their Acadian culture.

There was one group, however, that clung to the old ways and who consequently became a strong pole of attraction to those Magdaleners who had remained behind on the Islands, but who were discontent with their lot under the stewardship of John Fontana.

It started as a nucleus of a half-dozen families: Boudreaus, Landrys, Petit-Pas, Cormiers. They left the Islands intending to settle on Quebec's Middle North Shore, far from the Coffin regime but closer to the fishing banks they had been obliged to frequent since the American flotilla had driven them from their own waters. Stretching from the Natashquan Bank to the mouth of the Strait of Belle Isle, these banks were as familiar to them as the back of their own hand. From their new settlement on the North Shore, they also planned to profit from the enormous reserves of game on Anticosti Island, while still remaining within reach of the great herds of pagophiles that frequented the Gulf. Experienced navigators that they were, they thought it ought to be possible to approach the seal herd from the north, since the powerful tidal current of the Strait of Belle Isle prevents the formation of ice in the Jacques Cartier Passage, along the North Shore. It was a theory that remained only to be put to the test.

They left the island of Havre-aux-Maisons at the end of May 1857, on board the 40-ton fishing boat, *Mariner*, and spent the first part of June exploring the islands of the Mingan archipelago in search of a propitious place to settle. They were looking for a site with a safe harbour, arable land, ample reserves of wood, and a good supply of drinking water.

On June 10, after having carefully explored the entire region, the *Mariner* dropped anchor at Pointe-aux-Esquimaux (present-day Havre-Saint-Pierre), and the emigrants decided to settle there. They unloaded the animals, provisions, and tools they had managed to ferret away from the Islands and quickly set to work building cabins.

In October, the little group of exiles was joined by a new party of Magdaleners, led by Captain Christophe Vigneau of the schooner *Eugénie*. In March of the following year, after the *Mariner* had been wrecked without loss of life during the fishing season, the *Eugénie* left alone for the ice, carrying a crew of eleven men and five *canots*. It returned a month later with 620 seal pelts.

Although there was nothing exceptional about the size of this catch, it proved that the great herd of pagophiles could be reached as easily from Pointe-aux-Esquimaux as from the Magdalen Islands; and it encouraged many sceptics, discontent with living conditions in the archipelago but apprehensive of prospects for the future elsewhere, to pack up and move their households northward. The year 1858 saw the arrival in Pointe-aux-Esquimaux of at least a dozen more Magdalene families: Cormiers, Doyles, Jomphes, Arseneaus, Vigneaus. The following spring, five boats set out in pursuit of the seals. They returned with 2,021 pelts, worth more than $8,000.

Thereafter, the new colony drew more and more Magdaleners away from their native isles. Each year, dozens of families packed up and moved to the North Shore. They arrived in their fishing boats crammed amongst furniture, cattle, provisions, and tools, their hearts light in the knowledge that all this had been more or less purloined from the tyrant Fontana. There were joyful reunions; then work crews were organized, in which everyone participated, to construct lodgings for the newcomers. As a result they scarcely felt the pain of transition at all. Before they knew it, all traces of nostalgia had disappeared, and they began to look upon the magnificent crags of the Mingan archipelago with the same pride of ownership they had once bestowed upon the long golden beaches and the lovely red sandstone cliffs of the Magdalen Islands.

Nature was somehow more grandiose, more sweeping, more savage here than in the archipelago they had left far behind them. Perhaps it was the majestic encounter of land and sea, the sharp transition from the deep blue of the waves to the rich green of the forest. And the waters of the little rivers and streams descending from the highlands were filled with trout, muskelunge, and pike, as well as salmon in the spawning season. The earth, too, was reasonably fertile, though the growing season was short because of the latitude; but the unproductiveness of the soil was more than compensated for by the profusion of game and fur-bearing animals in the forest: hares, partridge, deer, moose – not to mention the flocks of teals, ducks, and geese that invaded the little islands of the Mingan archipelago during their migrations.

By the end of 1860, no less than thirty families (a total of 182 people) had emigrated from the Magdalen Islands to Pointe-aux-Esquimaux. Over the next five years, fifty-two more families followed in their wake. And the great exodus continued, though at a less frantic rate, until, by the end of the decade, there were 104 Magdalene families (more than five hundred people) settled on Quebec's Middle North Shore.

And there it ended. For some time, there had been serious concern on the Islands about the evacuation and its disastrous effects upon the local economy. An official inquiry was launched to determine the causes of the unrest and to suggest remedies for the situation – an investigation that resulted in the loss of more than a few feathers from the cap of John Fontana. About the same time, the Quebec government sent the schooner *La Canadienne* with its four canons to drive the American fishermen from the Magdalene fishing grounds. Its commander, Captain Pierre Fortin, acquitted himself of his task so well that he was elected, several years later, as the first member from the Magdalen Islands to the Quebec Parliament.

Living under a weakened, albeit tenacious, feudal regime that would not completely disappear until 1952, and able once again to profit from their rich fishing banks, the Magdaleners had no further reason to be dissatisfied with their lot and they remained at home.

The umbilical cord had been cut. Now on its own, the little colony at Pointe-aux-Esquimaux began to grow with the prodigious fecundity characteristic of the Acadian race. Before long, it was sending shoots all along the North Shore.

Thanks to the voluminous journal of one of its founders, Placide Vigneau, we know a great deal about the history, culture, lifestyle and accomplishments of this little colony, as well as the minor details of its day-to-day existence. Year after year, with an impeccable diligence, this historiographer kept a detailed record of all the salient events in the little community's existence, and as he enjoyed the double blessing of living to a ripe old age and remaining lucid right till the end, his work encompasses three-quarters of a century of the history of the Middle North Shore, from the arrival of the first Magdaleners in 1857 till his death in 1926.

This document is a real treasure![16] It offers us an intimate glimpse of the lifestyle of our ancestors: trapping for furs in the winter, hunting the seal in the spring, fishing for cod in the summer and herring in the autumn; salvaging the wrecks that periodically brought them a little unexpected bounty – ropes, sails, merchandise, perhaps even a fine copper compass; consorting with the *coureurs de bois*, the merchants and the adventurers – Jerseymen, Frenchmen, Canadians – who scoured the North Shore in search of exploitable natives or some lucrative bit of business; crossing the tempestuous December seas to Quebec or Halifax in their sturdy little boats to sell their catch at the highest possible prices; enjoying a profound knowledge of all things pertaining to nature and an unshakable faith in God.

From the beginning, the expatriate Magdaleners who settled in Point-aux-Esquimaux continued to hunt the seal, as they had traditionally done every spring. And as the population of the little colony grew, the sealing flotilla also expanded, with a proportional increase in the size of the catch. After 1870, more than twenty schooners were leaving for the ice each spring and returning with an average catch of some 6,000 pelts. This takes into account what Placide Vigneau calls the lean years like 1879, when the catch was a mere 782 pelts, as well as the fabulous springs of 1878, 1880, and 1881, when it rose to 11,091, 18,397, and 24,149 pelts respectively.

This trend continued for about twenty years, until the beginning of the 1890s, during which time the seal hunt accounted for half the annual revenue of the Pointe-aux-Esquimaux community, the other half coming from cod fishing. The pelt of an adult seal was worth a little less than a dollar and that of a whitecoat about half as much, and the oil fetched from 40 to 70 cents a gallon. Sealing, therefore, injected as much as $12,000 annually into the economy.

Then, around 1890, everything collapsed. Thereafter, no more than a few hundred pelts were harvested each spring. At the same time, the cod catch also showed signs of a serious decline, and the discouraged fishermen began to sell off their boats to the highest bidders. By the beginning of the twentieth century, the seal had all but vanished from the culture and economy of the Middle North Shore. The sea, once the sole benefactress of the local inhabitants, was no longer able to feed her own.

What exactly was the cause of this sudden ecological collapse? It is difficult to say, as there are very few clues upon which to base even an hypothesis. But judging from the evidence, it would seem that the great Newfoundland hunts were largely to blame for the catastrophe.

According to the reports of certain neglected or long-forgotten authors of the nineteenth century – Captain Pierre Fortin, Placide Vigneau, and others – harp seals were abundant in all parts of the Gulf until the 1870s. They were hunted as far downriver as Trois Pistoles. They swarmed over the ice in the early spring from Prince Edward Island to Anticosti Island, and all along the west coast of Newfoundland and in the approaches to the Strait of Belle Isle. These seals were apparently part of two distinct herds: one whose whelping ice was located in the west, the remnants of which are found today in the waters of Orphan Bank and Bradelle Bank; the other, located farther to the north and east, between Anticosti Island and the west coast of Newfoundland. The latter, which for all practical purposes is extinct, was the herd the Dorsets hunted; it returned to its former breeding grounds when climatic conditions proved favourable. In 1976, following a long moratorium on the industrial hunt in the Gulf, there were reports of an offshoot of the great herd comprising some 10,000 harp seals, drifting inexplicably from the north toward Cabot Strait, while the entire Gulf herd was still located west of the Magdalen Islands. This would seem to indicate a slight revival of the old Dorset herd.

Placide Vigneau, like any good fisherman who is careful not to divulge the location of his fishing grounds, rarely pinpoints the region in which the seals were captured. But judging from a few clues in his narrative, it would seem that the sealers of Pointe-aux-Esquimaux hunted a herd whose whelping ice lay somewhere between the eastern point of Anticosti Island and the western shores of Newfoundland: that is to say, the old Dorset herd. Geographically, this herd would have been in relatively close proximity to the great herd of the Front, separated from it only by the bottleneck of the Strait of Belle Isle; and since the two herds, while remaining distinct entities, were not completely sealed off from each other, it seems likely that the great Newfoundland hunts – and particularly the holocaust of 1884, from which not a single whitecoat escaped – had disastrous repercussions for the Dorset herd.

Other facts corroborate this conclusion. For one thing, the winter of 1884 was a long and severe one: the ice-pack didn't break up until very late that spring, and the fourteen miles of the Strait of Belle Isle were probably blocked with thick ice that prevented the passage of the seals. If so, the seals that would normally have whelped on the west side of the Great Northern Peninsula would have remained on the east that year, and their offspring would have been annihilated by the Newfoundland sealers. This is all the more plausible in light of Placide Vigneau's report that the approximately twenty-five boats from Pointe-aux-Esquimaux taking part in the hunt that year returned with a much smaller catch than usual: a mere 1,500 pelts. The explanation was that the seals were elsewhere!

The other corroborating fact is that the scarcity of seals suddenly began to manifest itself in the year 1890, or six years after the massacre. Now, the biological cycle of the seal – the time it takes to reach sexual maturity – is precisely six years!

Thus, the greed, the extravagance, the near-sighted rapacity of a handful of merchants unscrupulously exploiting large numbers of destitute people in the seal hunt, forced the Acadians of Pointe-aux-Esquimaux to abandon their lives on the high seas. Because they hunted the marine resources with wisdom, with that respect born of humility which is only nature's due, and because they never surpassed the modest measure of their means, these men were cast aside by the savage assaults of industry.

But the brief sealing history of the expatriated Magdaleners and their descendants will forever remain an example to all those who bear the name "sealer": in thirty years of sealing, pursuing the harps and hoods in their little sailing vessels across the icy, wind-swept waters of a boreal spring, not one man lost his life on the ice.

There were no widows or orphans in the Mingan archipelago who could blame the seals for their plight. Only a handful of boats, crushed by the ice or run aground on a reef, foundered during the hunt. Moreover, since these men always hunted in groups, the crew of a shipwrecked vessel could always find refuge on another boat. Occasionally, they would see the helpless wrecks of less fortunate vessels floating by, prisoners of the ice. Thus, in 1872, Etienne Landry of the *Marine* and Samuel Doyle of the *Vénélo* came upon the battered but still floating hull of the ill-fated brig, *Huntsmen*, out of Battle Harbour, Nfld., wedged tightly in the ice. Boarding it, they salvaged provisions, rifles, gunpowder, sails, and 320 seal pelts. The forty-two members of the crew who had abandoned the wreck several days earlier in an attempt to reach shore on foot all drowned at sea.

Superstition aside, there is a lesson to be learned from this: it is only when a man meets nature with deference and respect, meeting her on her own grounds and engaging her in a contest that pits the naked strength of one against the other, that she will treat him with clemency and provide him with his nourishment, without placing him in any undue peril. But when he tries to take her by force, with presumption and scorn, disdaining her majesty and her natural limits, she will confront him with savagery and avenge herself upon him.

This lesson applies as much to the Magdaleners as to anyone else. In the nearly two centuries of the inshore sealing history, the toll in human life has not once reached dramatic proportions. Of course, there have been tragedies, and they have left their mark upon the Islanders: they ooze from all the pores of local folklore. But if they seem so real, after so many years, it is precisely because they were the exception rather than the rule.

The greatest local disaster on the ice dates from about 1864, when two schooners, the *Gemma* and the *Breeze*, failed to return from the ice. Each carried ten crewmen: twenty sealers went to their deaths, leaving thirteen widows and forty-five orphans. In 1892, the schooner *l'Espérance* met the same terrible fate. But, to my knowledge, these are the only Magdalene schooners to go down with all hands while sealing in the Gulf – a mere trifle compared with the prodigious number of Newfoundland vessels that have disappeared in the ice of the Front.

Curiously, these two tragedies have been all but forgotten on the Islands today. They are commemorated by no song, no lament, no folk tale; and only a few old men still retain vague memories of them. Instead, it is the disasters of the inshore sealing campaign that are perpetuated in the rich oral literature of the archipelago.

An example is the gruesome tale of Eugène Turbide, the lighthouse keeper at Bird Rock.

On a calm, sunny day in March 1897, Eugène, his fifteen-year-old son and another young lad who had spent the winter with the Turbide family in their remote ice-bound retreat on Bird Rock, decided to go after the seals that were drifting past the lighthouse, a short distance out to sea. A lighthouse-keeper who spent his days watching the waves, Eugène was intimately familiar with the movements of the tides, and it is probable that he and the two boys hunted west of Bird Rock, so that later in the day the ebb-tide would carry them home.

It was a warm, balmy, spring day, the ice was firm, and the men worked in shirtsleeves, killing and sculping the whitecoats, intoxicated by the splendour of the icescape and the fresh salt sea air. But one must beware of fine days on the Magdalen Islands, especially at the approach of the high tides of the equinox, for just as surely as two and two make four, the gentlest calms invariably herald the most violent storms.

Hard at work, the sealers noticed too late that the tide had already begun to ebb. They had passed Bird Rock and were being pulled by the current toward the mouth of Cabot Strait.

Well, after all, this was only a minor inconvenience, and hardly the first time a party of hunters had had to spend a night on the ice! All they had to do was wait for the next rising tide to carry their floe back to the longitude of Bird Rock. The men were not dressed for a night in the open, but it was warm and calm, there were seals nearby, and Eugène probably profited from the occasion to show the young fellows how true sealers prepare for a night on the ice, wrapped up in sealskins.

Shortly after sunset, a wind came up out of the north and the temperature began to drop. The stars and the moon were hidden behind a bank of thick, black clouds, and the cold became intense. Then it began to snow. Eugène felt his optimism waning. The children whimpered and shivered in their light clothing, and Eugène held them in his arms to warm them. Then he ordered them to walk in circles, to walk and walk and not to stop, so as not to grow numb and freeze to death. But to no avail. By morning, the two boys were dead. Their frozen corpses lay on the ice, covered with a shroud of snow.

Eugène's mind reeled. He couldn't rid himself of the knowledge that it was he who was responsible for this tragedy: he had failed in his duty as a master watch. He wanted to die, but his robust constitution forbade him even that grace. For three days, he waited out the storm, his flesh black from frostbite, his limbs swollen and stiff, at death's door but unable to enter. On the afternoon of the third day, in a brief moment of lucidity, he saw that the wind had pushed his sheet of ice up to the foot of the high cliffs of Cape Breton Island, more than fifty miles from Bird Rock.

Back home, all too aware of what had happened to her men, Mme Turbide sat through those terrible hours alone. She lit the beacon-fire in the lighthouse to signal her distress. Sealers camped at Brion Island, some fifteen miles away, saw it, and knowing the fire was lit only during the navigation season, understood at once that something was amiss. On the first calm day, during a lull in the storm, they crossed to Bird Rock in a little ice *canot*, and learned the terrible news. With such a storm blowing, they knew there wasn't a chance in a million of the three castaways being found alive.

But Eugène Turbide *was* still alive. Fishermen from Pollett Cove, in Cape Breton, had picked him up on the ice and were looking after him as best they could. In his final hours, he began to rave, recounting the details of his terrible ordeal on the ice: the death of the boys, his long agony in the glacial tempest, his remorse, his implacable will to die. Finally, some ten days after his rescue, he gave up the ghost, having reached the limit of his physical and moral sufferings.

The memory of this drama was still fresh in the minds of the Magdaleners when, some fourteen years later, the Lebel tragedy occurred. This second catastrophe of the ice is commemorated by a lament[17] that is still sung today at wakes, its long slow rhythms and doleful melodies playing counterpoint to the whistling of the wind in the chimneys. Its depiction of the senseless and tragic drowning of six men corresponds in most respects with the memories a few old *Bassiniers* retain of the event.

On that day, the temperature changed abruptly and unexpectedly. About noon, the sky became choked with thick, black clouds and the wind suddenly began to cool. There were snow squalls and flurries of sleet, and folks back on land became concerned for the crews that had taken that morning to the ice.

By late afternoon, a crowd of spectators had gathered on the cliff-tops, watching for a sign of the returning sealers. But this was not the lively, loquacious, foot-stamping crowd that normally turned out to welcome the men back from the ice; it was a solemn, silent gathering that seemed almost to be crushed by the ominous atmosphere that lay over the day. As the minutes dragged on, the tension grew. No one spoke, but the dark, anxious looks that were periodically exchanged spoke more eloquently of the people's fears than any words could.

In the end, their apprehension proved unfounded. One by one, the crews came into sight: the Solomons, the Renauds, the Chevriers, the Gallants, the Jomphes, and the Lapierres appeared in the distance, zigzagging through the hummocks of the living ice, taking to the water, paddling across the mile-wide channel that separated them from the shore ice. Soon, all but one crew – that of Daniel Lebel – had been spotted. When the exhausted sealers arrived on land, covered with rime and frost, they assured one and all that there was no cause for alarm: the Lebels, they said, weren't far behind them.

And, indeed, about 4:00 PM, the Lebel party was spotted on the far side of the channel. All the anxiety and gloom that had hung over the day suddenly vanished to make way for an atmosphere of unrestrained euphoria. Once again, the Magdaleners had confronted adversity, the elements, danger, and had returned victorious. Even as they were congratulating one another, the Lebels took to the water, towing an enormous load of pelts behind them. The people went home to supper, knowing they had time to eat before returning to help the crew haul its *canot* to the top of the hill.

But when they had finished their meals and returned to the clifftop, the sea was empty. The *canot* and its men had vanished into thin air. No one had seen what had happened: disaster had struck slyly, in secret, at the least expected moment, sinking the boat and its six men. There followed an enormous outpouring of grief, underscored by a feeling of guilt, a sense of poignant remorse at having trusted fate when all the signs pointed to disaster. It is perhaps for this reason that the catastrophe has remained so indelibly engraved on the minds of the Magdaleners.

For these were not the only victims of the inshore sealing campaign. Other crews, other men – perhaps as many as a hundred in two centuries of sealing – ended their days out in the white kingdom of ice; but the details of their stories have all been forgotten. Their memories have become mingled with those of all the anonymous workers of the sea who have paid with their lives for some momentary imprudency or cruel caprice of nature.

It is a heavy toll for those who have lost a husband, a father or a friend. And yet, from another point of view, it's not such a heavy toll, in light of the dangers the men face every moment they are out on the ice. And this is a tribute to the inshore sealing campaign, when you consider the terrible catastrophes with which the history of the great industrial hunt of the Newfoundland Front is studded.

XIV
The Disasters

Of all the races of misers, extortionists, and usurers that have distinguished themselves throughout history, few have so completely embodied the concept of bloodthirstiness as the merchants of Newfoundland. Their scorn for human life – to say nothing of that of the seals – is equalled only by that of the ministers of the death industry itself, war. Next to them, Shylock seems a gentle, moderate creature, content with his simple pound of human flesh, while these merciless tyrants have sent it by the ton to rot in the vast communal grave of the ocean, simply to line their own pockets.

Even in the heroic days of sail, when the pagophiles were still abundant, the merchants reflected none of the humanitarianism, charity, and respect for one's fellow man preached by the Evangelists. But when the seals began to show signs of depletion and they felt threatened by the scarcity of resources, their alarm reached levels of pre-apocalyptic panic, reducing them to a state of utter monstrosity. In their inhuman eyes, the skin of the last sealer was less valuable than that of the last seal. So why spare one more than the other?

From the moment the seals became a rare commodity, the merchants' thirst for profits knew no bounds. They would have pelts and they would have them at any price! The markets were there, waiting, begging to be filled. It would be a crime, a barbarous act against nature, not to profit from them. The very reputation and prosperity of Newfoundland were at stake.

The results were disastrous. Though theoretically offering more security for their crews, at least a dozen of the new steam-powered vessels went down in the twenty-year period 1870-1890, with a loss of life that the merchants would doubtless have preferred to keep quiet. These vessels were called *Bloodhound, Wolf, Ariel, Retriever, Osprey, Hawk, Mic Mac, Tiger, Resolute, Windsor Lake* and *Mastiff*. One ship, the *Lion*, foundered in 1882, carrying nearly two hundred sealers to their deaths!

And these were not the only mishaps to occur on the ice: in 1874, the decrepit old furnace of the *Tigress* exploded, killing twenty men. In 1889, the *Walrus* lost thirty sealers, set adrift on an ice-floe that was never seen again. Five years later, tragedy struck the same vessel again when part of its cargo of dynamite, negligently stored in the crew's quarters, exploded, immolating twelve men!

So many disasters, occurring with such regularity, could not be ascribed to ill luck alone. They were the result of a pattern of systematic negligence, a flagrant scorn for the most elementary security regulations, and an increasingly savage campaign that was geared exclusively to the needs of the industry.

At that time, compulsory inspection of boats did not yet exist in Newfoundland. An owner could legally send any old vessel out to the ice, rigged out in any old way; and he never had difficulty recruiting hands and an ambitious captain to man these floating coffins. If a boat was no longer seaworthy, it was his opinion that the ice should seal its death warrant!

The captains, who became more and more competitive as the number of vessels and seals diminished, were prepared to take any risk to ensure themselves of a command the following spring. They knew they could expect no forbearance from the owners: if, through a lack of audacity or a simple concern for the lives of their men, they failed to return with a satisfactory catch, they could always be replaced at a moment's notice by any number of young bloods anxious to prove themselves on the ice.

In 1890, this unrestrained rivalry had almost catastrophic results. The *Terra Nova*, the *Eagle*, and two other steam-powered vessels reached the great herd just as the barometer was dropping sharply, a sure sign of unsettled weather ahead. Despite this warning, the captains ordered the sealers out onto the ice in groups of twenty-five or thirty, dropping them off over a large area of the ice-pack.

When the storm struck, with the suddenness of a lightning bolt, there were 650 men on the ice, and no means of rescuing them in the gusting snow and freezing rain that all but obliterated the icescape. Not a single boat made an effort to reach them. The mighty ocean swell was breaking up the ice, sending huge floes crashing against each other and filling the air with an outlandish din, and the captains deemed it more prudent to run for the open sea, where they stood a better chance of riding out the storm. Only one small boat, the *Kite*, less powerful than its rivals, failed to escape. Since its 150 crewmen were presumed to be in the immediate vicinity, its whistle was repeatedly sounded to indicate its position.

Fortunately, this little boat had a disproportionately loud voice for its size and the wind was blowing in the right direction. When they heard the shrill calls of its whistle, the 500-odd sealers of the other boats also headed towards it. That night and the following two days, the *Kite* played host to 650 sealers who otherwise would have been lost on the ice.

In 1898, tragedy struck. It was near the end of the sealing campaign. The boats were almost all loaded with fat, and they might already have been on their way back to St. John's if a few of them hadn't decided to compete, right to the end, for the Highliner flag awarded each year to the captain who returned with the largest catch.

The SS *Greenland*, its hold full of pelts, was one of these vessels.

On the morning of March 31, despite an alarming drop in the barometer, the captain of the *Greenland*, Arch Davis, ordered his 150 sealers out onto the ice. It was a mild, balmy day, and most of the men set out in shirtsleeves. There remained only a few isolated patches of seals spread out over the ice, and they knew they would have to walk a long way to reach them.

In the early afternoon, the wind suddenly shifted out of the north-west in frightful gusts. One of these caused the *Greenland* to list so badly that its entire cargo of pelts, thrown every which way in the open hold, shifted to one side and left the vessel tilting at a precarious angle. Fortunately, part of the crew was not far off and returned to correct the situation.

Then, as invariably happens with these strong nor'westers, the temperature suddenly began to drop. In a few hours, it fell nearly 25°C. The spring-like morning was followed by a Siberian afternoon with temperatures as low as −20°C, and signs of an even bleaker night ahead. Meanwhile, the crew of the *Greenland*, busy redistributing the cargo in the ship's hold, had failed to keep an eye on the movements of the ice. Before they knew it they were trapped, imprisoned in a small pond encircled by high ridges of ice. There was now no question of the vessel leaving in search of the more than one hundred sealers still on the ice.

Night fell, thick with shadows and drifting snow. The wind wailed in the shrouds like an interminable death cry. The sinister creaking of the hull as it scraped against the ice made sleep impossible. Even in the forecastle, where a small coal fire burned fitfully, the cold was all but unbearable.

In the morning, during a brief lull in the storm, the man on watch in the crow's nest sighted a group of sealers about two miles off, huddled together for protection against the cold. There were about fifty of them, most in a pitiful state. They were immediately brought back on board. The ship's doctor, who for reasons of economy was only a young apprentice pharmacist, had to amputate a number of frozen limbs in cold blood: a shot of rum and a bullet between the teeth (to keep the patient from biting off his tongue) was all the anaesthetic available for this butchery!

Throughout the day of April 1, the *Greenland* remained imprisoned in the ice, and it wasn't until the following afternoon that it managed to break free to go in search of the 52 men still missing. Twenty-nine of them were discovered shortly before nightfall: a little group of corpses that, in the waning light, bore a grotesque resemblance to a mausoleum of sculpted figures seized by death in the most incongruous of postures. Blocks and tackles had to be used to lift these statues of petrified flesh, black with frostbite beneath their thin cotton shirts, onto the vessel: prostrate men encased in shells of ice from which they were removed with picks; kneeling men, their hands tightly clasped, imploring a final forgiveness; a father holding his two sons in his arms more inseparable in death than in life.

Miraculously, a few of these men were still alive, but they were in such a sorry state that all but four died on board ship. The bodies of the other twenty-three victims were never recovered.

Back in St. John's, a crowd of dumbfounded spectators witnessed the return of the *Greenland* and its macabre cargo of corpses with a rising sense of nausea and indignation. Massed on the wharf, they watched the army of lame, amputated, and mutilated being unloaded and rushed to hospital; then they swarmed over the bridge of the vessel to identify the corpses, which were lashed to the mainmast.

They received from the witnesses a first-hand account of the exact circumstances of the catastrophe; and for the first time since men had begun to die on the ice, they rebelled. The forty-eight victims of the *Greenland* were the last straw. That men were sent out onto the ice to risk their lives in search of seals was tolerable, but that they should be hastened to their deaths by sheer vanity, when the hold was already full of blubber, was surely insufferable!

For several years, the fishermen of Newfoundland had been trying to organize themselves into a syndicate, known officially as the Fishermen's Protective Union. The tragedy of the *Greenland* was to provide the stimulus needed to allow this still embryonic organization to be born. It galvanized public indignation into a vast protest movement that cried out for a public inquiry into the disaster, so that charges could be laid against the guilty parties.

This time, the sharks of Water Street did not get away with the empty messages of condolences they habitually issued following a catastrophe on the ice – assuring the public that the lost ones were safely in heaven under the watchful gaze of the Almighty, where they were so much better off than they had ever been back on earth. Nor did the few lines they published in the newspapers in honour of "the brave and courageous soldiers of the industrial army, who fought, suffered, and perished on the icy battlefields for the Captains of Industry and their loved ones, RIP," appease those who cried out for justice.

Sensing that they were truly in hot water this time, the merchants resorted to a final stratagem: they organized a collection on behalf of the widows and orphans of the victims of the *Greenland*, and managed to collect a grand total of $400 from the good souls of St. John's, without even having to loosen their own pursestrings. But the hypocrisy of this gesture did not go unnoticed, and pressure continued to mount on the government until it finally consented to hold an inquiry.

There was just one hitch: the merchants had the government – along with just about everything else in Newfoundland – under their thumb, and they saw to it that the statements made before the judges did not compromise their reputations. As a result, the inquiry proved to be a monumental farce: those who had escaped the disaster were not allowed to testify; only the captains and the shipowners were heard. And, far from leading to judicial proceedings, the hearings resulted in a general acquittal of the responsible parties. The *Greenland* tragedy, it was concluded, had been an "act of God," and the case was closed.

But the affair was not forgotten. Public rage continued to smoulder in silence, flaring up periodically in bitter disputes between exploiters and exploited. The unfortunate conclusions of the inquiry had merely widened the rift between these two classes, and added fuel to the cause of The Fishermen's Protective Union, which now had proof of the collusion between the legislative, executive, and judiciary branches of the government. A literate fisherman by the name of William Coaker founded a newspaper entitled *The Fishermen's Advocate*, which served as a catalyst for stirring up public indignation; and it wasn't long before The Fishermen's Protective Union moved into the political field and began to elect members to the House of Assembly. By 1915, following another major tragedy of the ice, it had elected enough members to hold the balance of power.

The representatives of The Fishermen's Protective Union attended debates in the Assembly in rubber boots and heavy woollen sweaters impregnated with the odour of fish and brine. No one wanted to sit beside them. Their very presence in the hallowed precincts of the House of Assembly united the traditional political foes of British-inspired regimes, the Whigs and the Tories, who saw them as dangerous mutineers aboard the great vessel of Newfoundland.

Their mutiny, however, was a peaceful one. They did nothing but demand stricter security measures on the boats, more humane treatment of the men, and higher remuneration for their work; and it was rare that they went so far as to call a strike. When this became necessary, as in the case of the famous strike of 1902, it took only the smallest concession on the part of the merchants to restore order.

That year, on March 10, the day the steam-powered vessels were scheduled to leave St. John's for the ice, the 300-odd crewmen refused at the last moment to embark. Forming an impressive cortège, they paraded down Water Street to the Parliament Buildings to present their grievances to the Governor in person. Their demands were fourfold: first, that the price of blubber be raised from $3 to $5 a hundredweight; second, that the policy of renting berths at $3 each be done away with; third, that the recent 35 per cent increase on all materials sold on board the vessels be abolished; and, fourth, that the owners be obliged to feed the men while they were outfitting the ships for the ice.

The strike lasted two days, then a compromise was reached, attesting – as one commentator put it – to "the wisdom of the capitalists and the magnanimity of the sealers." Oil was thrown on the troubled waters: the owners agreed to do away with the rental of berths and guaranteed a price of $3.50 per hundredweight of fat. Appeased, the sealers returned to their boats and the vessels departed at once for the ice.

In the House, the fiercest adversary of The Fishermen's Protective Union was Abraham Kean, Minister of Fisheries in the Conservative government. A legendary sealing captain with a long list of honours (by his death in 1934, he had brought back more than a million pelts!), feared more than admired for his reputation as a pitiless slave-driver, he staunchly opposed all improvements in the lot of the sealers, maintaining that crews too comfortably lodged and too well fed were inclined to be lazy and disobedient.

He was a big, robust man, with short stubby fingers and a cold, piercing gaze that could cause the boldest man to lower his eyes. A sober, pious individual, with a full, neatly trimmed beard, he seemed faultless on the surface, though it was murmured here and there that he was not averse to helping himself to another boat's pelts if the vessel was well out of sight, and that he took undue advantage of his position as Commodore-of-the-Sealing-Fleet-cum-Minister-of-Fisheries to procure commands for his sons.

In response to what he considered the fallacious and irresponsible nature of the reforms demanded by the Fishermen's Party, Abraham Kean stressed the hard-line policy – the only approach which, in his opinion, would bring prosperity and stability to the island. In this climate of conflicting philosophies, the stage was set for a new tragedy.

It struck in 1914, and proved to be the worst on record.

By this time, the face of the Newfoundland sealing fleet had undergone a great many changes. The old "wooden walls," originally brought back from Scotland, had practically all disappeared – eleven of them had perished on the ice between 1907 and 1912! – and had been replaced by steel-hulled vessels. The introduction of these solid, powerful ships was necessitated by the fact that the sealers had to push farther and farther onto the icefield each year to reach the seals; the herd that had once covered the entire pack was now confined to a small area in its centre.

The competition for these steel-hulled vessels had prompted enormous bids from St. John's three largest merchants. Harvey & Co. started the ball rolling in 1905 with their acquisition of the SS *Adventure*, an 880-ton vessel with a 200-horsepower motor, whose performance on the ice proved exemplary from the start. Thus encouraged, the same firm immediately placed orders for two additional 500-ton, 300-horsepower vessels, the *Bonaventure* and the *Bellaventure*. Bowring Brothers countered by ordering the 2,000-ton, 450-horsepower *Florizel*, a veritable monster; and not to be outdone, Job Brothers & Co. commissioned the construction of two vessels: the *Beothic*, a vessel similar to the SS *Adventure*, and the 1,000-ton ice-breaker, *Nascopie*. Stung to the quick, Bowring quickly outbid its rivals for the super-giant of the fleet, the *Stephano*, a 2,100-ton vessel with a 600-horsepower motor. And there the bidding ended: this was 1912 and, crafty businessmen that they were, the merchants knew there did not remain enough seals to justify any further major investments.

The year 1914 augured well for the sealing industry. There were rumours of a war in Europe, with the possibility of a spectacular increase in oil prices, and the merchants were not going to let such a lucrative opportunity escape them. Consequently, they equipped all the boats they could lay their hands on for the ice. The less decrepit old whaling boats and the steel-hulled vessels were sent to the Front; while the remainder of the fleet, as sorry a collection of floating wrecks as has ever been seen, were dispatched to the Gulf, where the ice-pack is less murderous than out in the Atlantic. There was only one flaw in their reasoning: the journey from St. John's to the Gulf ice and back again involves a circuit of more than five hundred nautical miles through the treacherous seas of the Cabot Strait, at that time of the year one of the most dangerous bodies of water in the northern hemisphere! The old galleys, rotten and leaking at every seam, were clearly not up to such a journey.

As usual, the Kean family was well represented on the ice that year. Abraham, the Commodore of the Fleet, commanded the *Stephano*; his eldest son, Joe, the *Florizel*; and one of his younger sons, Wes, the *Newfoundland*, the largest and most solid of the old wooden vessels. Of all the boats sent to the Front, this was the only one not equipped with a wireless or a barometer – its parsimonious owner, Alick Harvey, being of the opinion that these instruments were an unnecessary luxury on such a decrepit old boat. Instead, he had had them removed and installed on one of his new steel-hulled vessels.

Wes Kean, moreover, was only twenty-nine years of age, and possessed neither the qualifications nor the experience required of a sealing captain. He had obtained his command, three years earlier, due to his father's influence; but as he had discharged his duties honorably during his first two springs out on the ice, wisely following in his father's wake and scrupulously obeying his orders, a captain he remained.

This year, however, with no means of communicating with the other vessels in the fleet, his luck ran out. From the very beginning, he remained imprisoned in a formidable ice-jam, far from the seals, and all his efforts to extricate himself were in vain.

On March 29, while Captain Billy Winsor of the *Beothic* was preparing to return to St. John's, with 20,000 pelts in his hold and an additional 5,000 stacked on the bridge of his vessel; while his father, Abraham, had packed 18,000 pelts into the enormous hold of the *Stephano* and was furiously continuing the hunt in the hope of surpassing Billy Winsor's catch; and while his brother, Joe, and the other captains had each harvested between 10,000 and 15,000 pelts, Wes Kean hadn't slaughtered a single seal and was fuming with impotent rage.

Without a telegraph, there was no way he could know the size of the other boats' catches; but judging from what he had had ample leisure to observe through his binoculars – the slowly moving columns of black smoke on the horizon, almost at ice level, and the spouts of white vapour indicating the use of windlasses – he had no doubt that the season was shaping up well. What he didn't know, however, was that several days earlier, on March 23, tragedy had almost struck.

It was the classic scene. Captain Billy Winsor had set a party of fifty-odd men down on the ice, then moved on to pick up the pelts panned by his other teams some fifteen miles to the west, afraid perhaps that the *Stephano*, prowling in the vicinity, might make off with them. Meanwhile, a strong wind had come up out of the north, causing the channel leading to the west end of the ice-pack to close up, and the men had had to spend the night on the ice. They had built themselves a shelter with blocks of ice and seal carcasses and had kept themselves warm over a fire of seal fat, sitting on live whitecoats to keep their backsides from freezing. That might have been the end of it if, the following morning, returning to their boat, the men hadn't informed their captain that the previous evening they had observed a large horned beast snickering at them through the cloud of dense snow – a sure sign of disaster.

Announcing the rescue of his men over the radio, Billy Winsor informed the other captains of this strange apparition. In his mind, it was clearly an evil omen, and from that day forward he did not allow his crew to venture more than a mile or two from the *Beothic*.

On board the *Stephano*, Abraham Kean scoffed at this news: the Commodore of the Sealing Fleet was not about to take such nonsense seriously! Those men had merely been hallucinating; what they had witnessed was no more than a projection of their own fear and helplessness – qualities for which they were well known. Listen to them and not another seal would be slaughtered that year on the eastern shores. And then the seals would destroy the fish – that's what would happen!

Leaning on the rails of the *Newfoundland*, unaware of the *Beothic*'s near-tragedy, young Captain Wes Kean was cooling his heels on the evening of March 29, when he thought he perceived, in the scarlet glow of the setting sun, the stately silhouette of the *Stephano* floating between sea and sky, about a half-dozen miles away. After conclusively identifying it through his binoculars, he decided to send his men in that direction first thing the next morning.

Long before dawn, on Tuesday, March 30, the 176 sealers of the *Newfoundland* leaped onto the ice, with orders to make their way on foot to the *Stephano* and to report to Captain Abraham Kean, who would take them to the seals.

It was a fine day – too fine, in fact, for those who recalled the *Greenland* disaster, sixteen years earlier almost to the day. The stars were pale; not a breath of wind troubled the ice-field. From deep in the fissures and crevasses could be heard the crystalline sounds of the frosy. The sealers, anticipating a warm day and a long walk, set out in their shirtsleeves.

If the young captain's vessel had been equipped with a barometer, he would have been suspicious of this unusual calm – not to mention the mirage-like appearance of the *Stephano* the previous evening, when the boat was in fact too far away to be clearly seen – for the atmospheric pressure had been falling at a dizzying rate all night.

About three hours after they had set out, the men came upon a patch of about a hundred seals, and they stopped to slaughter them and to invigorate themselves with the ritual mouthful of hot blood. Then they panned the pelts, marking them with their boat's flag, and were about to move on when they were suddenly caressed by a tepid, spring-like breeze. It came out of the south-east, blowing in soft, warm gusts, as if the sky itself were breathing. To the great astonishment of the others, about fifty of the men expressed a sudden desire to return to the boat, explaining that these were the first signs of a cyclone.

The remainder of the crew scoffed: they called them curs, shirkers, cowards; they showered them with scorn and abuse. The master watch, George Tuff, who at the age of seventeen had been one of the few survivors of the *Greenland* tragedy, threatened to dock their pay if they didn't remain with the main body of the crew. But not even this would deter them: if the others didn't want to follow them, that was their right, but they were returning to the *Newfoundland*.

By the time the remaining 126 sealers reached the *Stephano*, six hours later, it was clear that the "cowards" had not been wrong. Foul weather was setting in quickly. The wind was blowing vigorously out of the south-east and large flakes of wet, heavy snow had started to fall. It was with a keen sense of relief that the crew of the *Newfoundland* boarded Abraham Kean's boat, knowing they could ride out the storm safely there. While George Tuff went off to explain the situation to the "old man" and to receive his orders, the men went below to the forecastle, where they were served bowls of hot tea.

As the *Stephano*'s crew was still out on the ice, the men from the *Newfoundland* were not surprised to hear the 600-horsepower motor of the big vessel suddenly come to life: Abraham Kean must be on his way to pick up his sealers. It would be necessary to double up in the crew's quarters that night, but that was all right; they would be all that much warmer, since they had nothing but cotton shirts on their backs.

The *Stephano* steamed along for about half an hour, then came to an abrupt stop. From the bridge, the men heard the sound of Abraham Kean's stentorian voice ordering them onto the ice.

Arriving on deck, the men saw that the storm had set in for good: thick snow was falling, whipped by a glacial wind that whistled shrilly in the shrouds. Presuming that Abraham Kean wanted them to work a little patch of seals in the immediate vicinity or to retrieve a pan of pelts, they leaped onto the ice; but they began to have serious doubts when, a few moments later, the *Stephano* made a half turn and disappeared, slipping into the thick fog of snow that rendered visibility almost nil.

One hundred and twenty-five pairs of bewildered eyes turned toward George Tuff.

"Okay, byes," he said simply. "Our orders is to work a big patch of seals about a mile from here, then to make our way back to our own ship."

Every man knew that this was insanity. It was almost four o'clock in the afternoon, and even if they located the seals without delay, which was unlikely in this raging blizzard, they wouldn't finish working them before six. Then, they would still have to walk at least a dozen miles back to the *Newfoundland*. It would be early morning before they arrived! But their fear of disobeying an order was even stronger than their fear of death; they searched until dark for the seals without finding them, then, in the blinding snowstorm, they set out resolutely in the direction of their ship.

The storm, which had only just struck the Front, was to prove one of the worst in memory. On Tuesday, March 30, it buried the south-east section of the island of Newfoundland beneath more than fifty centimetres of snow, unleashing winds that lifted roofs from houses and toppled barns.

Farther to the west, a little earlier in the day, it had surprised the fleet of decrepit old "wooden walls" chugging back from the Gulf, their holds filled with blubber and pelts. Aware of their limitations, these ancient vessels had run for shelter in the deep bays along the southern shores of Newfoundland – all but one, the *Southern Cross*, which had decided to ride out the storm. Its holds filled and its decks piled to the gunwales with pelts, its captain coveted the honour of being the first boat to return to St. John's with a full load of fat. Surely this was reason enough to risk his life and the lives of his 173 crewmen.

Its engine gasping, the old galley laboured through the turbulent seas of the Cabot Strait, stubbornly riding the waves. It plunged heavily into a yawning trough, its decks submerged to the wheelhouse in churning green waters, its screw spinning madly in the air; then, rolling on its side, it reared up on another billow, floods of water streaming from its decks, its prow pointed straight to the sky. It hung there for a moment, suspended between sea and sky, before being caught by a powerful gust of wind that tipped it on its other side, its masts level with the water, and sent it plunging back into another trough. Thick clouds of black smoke spewed from its chimney, as it strained at the seams, threatening at any moment to disappear forever beneath the waves; but each time it rose again, sliding phantom-like through the whirling snow and spindrift, riding the breakers, while the entire crew took turns at the wheel and manned the pumps to keep the vessel afloat.

Late in the afternoon, the *Southern Cross* passed Cape Race: the operator of the Marconi station sighted it from the top of the lighthouse during a lull in the storm. It was dangerously close to the reefs, its massive prow ploughing the mountains of seawater, looking more like a submarine than a surface vessel. And that was the last that was seen of it, with its crew of 174 sealers.

Back in St. John's, they waited weeks, months, with the mad hope of widows and orphans of the sea that the wrecked boat might only have gone astray, might have run aground on some distant, deserted isle from which it was impossible for the crew either to return or to give a sign of life. But, in August of that year, they were obliged to acknowledge the facts: several planks from the prow of the lost vessel were found washed up on a beach in Ireland, smashed almost beyond recognition but bearing the half-effaced inscription: . . *uthern C* . . . . . It was as if the remnants of this Scottish-built whaler had wanted to return to their native shores to bear witness to the miseries and torments they had experienced on the far side of the Atlantic.

The storm that sank the *Southern Cross*, probably during the night of March 30-31, moved northward and gained in intensity as it approached the ice of the Front.

On board the *Newfoundland* – stripped of its wireless – Captain Wes Kean had blown out his candle early that evening after a good hot meal, secure in the knowledge that the 126 men he had sent out on the ice earlier in the day were now safe and sound on his father's vessel. And his father, Abraham, slept just as soundly in his cabin on the *Stephano*, certain that by now the crew of the *Newfoundland* would have made their way back to their ship, as he had ordered them to do. In his mind, it was inconceivable that any order of his would not be carried out to the letter.

It was the men of the *Newfoundland* who paid for this blunder as they groped blindly through the darkness, the snow swirling about them. Not only were they in trouble, but no one was aware of it. The boats' whistles, which normally would have been sounding to indicate their position to them, were silent.

Buffeted by the wind and the mighty ocean swell, the condition of the ice deteriorated hourly. The pack broke up into smaller and smaller fragments, interspersed with vast streams of slob ice and frosy impossible to identify beneath the thick layer of snow. Before George Tuff called a halt, five men had already "passed through" and could not be rescued.

The entire *Greenland* tragedy, in all its stark horror, was being repeated. Since walking had become too perilous, the men camped on a large ice-floe to pass what remained of the night. They built a wall of ice blocks, behind which they took refuge, and they lit a pitiful little fire that quickly reduced their gaffs and flag-poles to ashes. Then they had to jump up and down, walk in circles, run, wear themselves out in one way or another to keep warm, for most of them were in shirtsleeves.

Meanwhile, the wind veered around to the east and the snow changed to freezing rain. At dawn, cold, dry, gale-force winds blew out of the north, transforming the wet clothing of the sealers into shells of ice.

Besides the bowl of tea they had drunk on board the *Stephano*, the men had eaten nothing for nearly twenty-four hours. Exhausted, the weaker of them lay down on the ice to die. Others took leave of their senses: laughing hysterically, they threw themselves into the water and drowned.

The stronger men continued to move about, clinging desperately to life, while the cold became more and more intense. About noon of March 31, the snow let up, and from the height of a pinnacle of ice, George Tuff spotted the boats several miles away. He at once ordered his crew to begin walking in that direction.

Two hours later, despite gusting snow and the precarious state of the ice, the hundred or so survivors, exhausted, frost-bitten, found themselves almost within hailing distance of what appeared to be their salvation. Climbing another hummock, George Tuff spied the *Bellaventure* less than two miles away; and she seemed to be heading straight in their direction!

There was a moment of euphoria, a burst of hope! The men took off helter-skelter, running as fast as their frozen legs would carry them, waving their arms wildly in the air, and wailing like banshees. It was the *Bellaventure*, no question about it, and the look-out must have seen them, for she was still moving in their direction! Now, there was no more than a mile separating them!

But like the other vessels in the fleet, the *Bellaventure* was unaware that the *Newfoundland*'s men were out on the ice. She was simply cruising the waters, with all her crew on board, inspecting the changes wrought by the storm upon the ice-floes. Less than half a mile from the men, who were all but invisible in the drifting snow that masked them to the height of their shoulders, she came about and promptly showed them her stern.

This was the last straw for many of the men. Famished, perishing with cold in their light clothing, their hands and feet as hard as rocks, they preferred to give themselves up to the comfort of death than to prolong their ordeal. The remainder of the ragged crew moved on, putting one foot mechanically in front of the other, driven more by blind reflex than the conviction that they would ever reach safety. The men clung to each other to avoid falling, stumbling on the smallest obstacles. Little by little, they inched their way towards the boats that were just visible in the distance, steaming aimlessly to and fro in the ice – tiny mirage-like figures with great plumes of black smoke nodding gently above their heads. It was their only hope.

Once again, their efforts came close to being rewarded. Late in the afternoon, the wind died down and they spotted the *Newfoundland*, no more than four miles away. Thanks to the break-up of the ice-field in the storm, it had finally freed itself from the ice-jam and was steaming leisurely in the direction of the *Stephano* to pick up the missing members of its crew.

Taking about ten of his most able-bodied men, George Tuff set off in its direction. If they hurried, they could cover a mile or two before nightfall; that should be far enough to be seen by someone on board the vessel. The weaker sealers were ordered to wait; but, all but driven mad by their long ordeal, the men thought he simply wanted to abandon them there on the ice and they tried to follow him, some on their knees, some on all fours, some crawling flat on their bellies, uttering bitter reproaches in their failing voices. Never had the ice-floes been the scene of such a grotesque and pitiful procession.

When the scouts judged they had gone far enough to be seen from the boat, which was once again caught in the malevolent ice, they began to wave their arms frantically, like disjointed marionettes, making gestures that would have been spotted if anyone on board had been looking in their direction. Unfortunately, the crew was all on the farther side of the vessel, gazing in the direction of the *Stephano*, wondering what was keeping their mates from leaping over the railings and heading back to their ship. It was unusual for the "philanthropic" Abraham Kean to offer hospitality to another ship's crew for any longer than was absolutely necessary.

Night fell, colder than the heart of a Water Street merchant. The mercury dropped to $-20°C$, and the wind blew fiendishly out of the north, intensifying the already unbearable cold. It was more than most of the men could endure. One by one, they succumbed. Yet, strangely, while most of them had long ago resigned themselves to death, they continued to resist it; as if the very fact of having struggled for so long made their fate all the more unjust and abominable.

One young lad, unwilling to see his cousin die, began to jump up and down on his frozen legs in an effort to rouse him; but all he did was shatter the poor man's legs and finish him off. Some men, expiring from hunger and thirst after forty-eight hours of total abstinence, bled to death after slashing their wrists to drink their own blood. The dead were stripped of their clothing to provide the living with warmth; there was even some talk of eating the corpses.

By Thursday dawn, the wind dropped a little and the visibility was much improved. Climbing to the crow's nest, young Captain Wes Kean trained his binoculars in the direction of the *Stephano*, several miles away, wondering what was keeping his men from heading back to their ship; then he shifted his gaze in the opposite direction, hoping perhaps to spot some seals. In the pale early morning light, he caught sight of a group of men, more dead than alive, dragging themselves in the direction of the *Newfoundland* and he grasped the full horror of the situation, the monstrous misunderstanding that had allowed him to presume that his 126 sealers had found refuge on his father's boat while, in fact, they had been out there on the ice for two days, without food or supplies, in the worst storm in recent memory!

He almost broke his neck scrambling down the ladder, yelling frantically at his men to go to the rescue of their unfortunate crewmates. Half an hour later, George Tuff and three of his companions were lifted on board, frozen stiff, their hands and faces black from frostbite, their clothing hard as a rock.

At about the same time, on that morning of April 1, the *Bellaventure*, skirting the ice-floe in search of seals, came upon a group of twenty-odd sealers, presumed at first to be a hunting party from another boat. But, approaching them, Captain Robert Randell was struck by their unusual behaviour: they were staggering about, reeling like drunkards, some of them on their hands and knees, one or two crawling flat on their bellies.

He immediately sent men to their rescue, and a short while later, the wireless tapped out the terrible news to the other boats in the fleet:

"Captain of *Bellaventure* to Captain of *Stephano* stop twenty men from *Newfoundland* picked up this morning in very bad shape stop on ice since Tuesday morning stop many others dead stop . . ."

At once, rescue crews from all the boats, carrying food, rum, and stretchers, began to scour the ice for survivors and corpses. By evening, forty-five mutilated men, most of them crippled for life, had been brought on board the *Bellaventure*, along with sixty-two corpses. The bodies of the remaining fifteen victims of the disaster were never recovered.

Back in St. John's, the news of the *Newfoundland* tragedy, transmitted by wireless at lightning speed, stunned a public that was already anxious about the delay of the *Southern Cross*, last sighted two days earlier as it passed Cape Race, less than a half-day's run from its destination. Though they had not given up hope of seeing the old "wooden wall" again, no one could help but fear the worst; and, mentally adding 174 to 77, even the most optimistic observer had to admit that only a miracle could prevent 1914 from going down as the most disastrous year in the history of the sealing campaign.

The only people apparently not moved by this sombre prospect were the merchants. If they were troubled at all by the dispatches that came in with reports of the *Newfoundland* disaster – and it is not at all certain that they were – they took great care to conceal it. With an astonishing display of *sang-froid*, they instructed the captains of their ships to continue the hunt. Only the *Bellaventure* was permitted to return with its cargo of pelts, cripples, and corpses. It would take more than an unfortunate accident – which, needless to say, they deplored as much as anyone – to interfere with the hunt; especially in that glorious year of 1914, when the market for oil promised to be more lucrative than ever. Newfoundland could not afford to turn up its nose at such an opportunity.

Their decision was not well received, either by the public or the sealers. Had they spat in the faces of the victims, it would not have been more scandalous. For the first time the crews of the sealing vessels mutinied at sea.

On the morning of April 2, when Captain Billy Winsor of the *Beothic* ordered his men overboard onto the ice, they categorically refused. They weren't going to set foot in that graveyard again, not if the King of England himself ordered them to! Since the hold was already filled with a record catch of more than 25,000 pelts, the captain did not insist, and pointed his prow toward home.

Then the mutiny spread to other boats. On the *Bloodhound* (the second vessel to bear that name), the sealers, who had had only three hot meals since their departure for the ice, seized their captain and forced him to return. On the *Nascopie* – (where the publisher of the *Fishermen's Advocate*, William Coaker, wrote angrily in his journal of "the heartless lovers of gold who reap the cream of the seal fishery without sharing its dangers and without a trace of respect for those who risk their lives from year to year to maintain them in luxury"), the sealers simply failed to appear on deck when the captain ordered them up.

On the *Florizel*, the *Adventure*, the *Diana*, the *Kite*, the *Terra Nova*, and all the other vessels of the fleet, similar acts of insubordination occurred; and, one by one, the boats came about and headed for home. Only the *Stephano* remained at the Front: Abraham Kean was not one to bow before resistance. He armed his officers and ordered them to defend the bridge against any act of sabotage. Having already sent seventy-seven sealers to their deaths, he was prepared to deal similarly with anyone who demanded that some respect be shown their memories.

Several days later, realizing that his efforts at intimidation had failed, he opted for a paternalistic approach and called the crew on deck to negotiate. Mark Sheppard, the leader of the mutineers, didn't give him a chance to make any great speeches. He moved directly to Kean and, in a voice loud enough to be heard by everyone, stated the crew's reason for refusing to return to the ice: "After what I seen of this disaster through your neglect, I don't think you're competent to look after men."

It took the Commodore of the Fleet several days to swallow this insult, but seeing that his crew was resolved not to set foot back on the ice – even when he had the impertinent Sheppard put in irons and threatened to dock their pay – he gave in and turned back to St. John's.

It was April 8, and a storm of indignation was fulminating against him throughout Newfoundland.

XV
The Great Seal Herd

It was over.

Times were changing, in Newfoundland as well as in the rest of the world. Thanks to recent improvements in communication, the dual disaster of the *Newfoundland* and the *Southern Cross* made headlines, not only on the island, but across Canada, the USA and as far away as Britain, where the Governor of Newfoundland, Sir Edward Morris, was on an official visit. The whole thing was an acute embarrassment to him, for the press insinuated that the two disasters might have been prevented if stricter measures had been adopted to ensure the safety of the men in the perilous adventure of sealing.

To clear up any doubts as to who was responsible for the catastrophe, the government set up a commission of inquiry. Conscious that the eyes of the world were on it, it was obliged this time to admit the testimony of the survivors, and to give them an equal hearing with the captains and owners of the fleet. It was no easy business for the powerful establishment of Water Street to be confronted publicly with these crippled, mutilated men, these amputees who followed one another to the stand, some on crutches, some flat on stretchers, exhibiting their blackened flesh and their scars for all to see, muttering bitter reproaches in their broken voices.

Alick Harvey was asked what had possessed him to remove the wireless and the barometer from the *Newfoundland*. He calmly replied that he had decided it was a waste of money to keep these instruments on such an old vessel when they could be of more use elsewhere. And when he was asked if, in reaching this decision, he had considered the safety of the men, he replied bluntly, "Not at all."

As the inquiry proceeded, it became increasingly evident that it must end in laying charges of either involuntary manslaughter or criminal negligence. But the testimony of the Minister of Fisheries, Abraham Kean himself, completely altered the face of the hearings: singlehandedly, he rescued the sharks of Water Street from their tight corner. In a theatrical tirade resonant with an impassioned patriotism, the very man who had sent the crew of the *Newfoundland* to their deaths declared that he had been acting in the interest of the sealers, and at the same time of the entire Newfoundland sealing industry, in ordering those men onto the ice. Was it his fault if they had failed to obey his order to return to their ship? Should he be held responsible for their insubordination?

It didn't take much more of this line of reasoning for the commission to decree once again that the disaster had been an "act of God," for which no human being could rightly be held responsible. However, it did recommend that the House of Assembly take immediate steps to legislate security regulations governing the annual seal hunt. These, it stressed, must include the installation of radio equipment on all vessels and the annual inspection of boats. The case against the merchants was dismissed.

But times *were* changing. The following spring, the inspectors came down hard on the shipowners: only three "wooden walls" of more than a score were deemed seaworthy; the remainder had to undergo considerable costly repair before being allowed to leave for the ice.

In the House of Assembly, where The Fishermen's Protective Union now held the balance of power, the days were gone when the ruling party could get by with merely amending the laws governing the seal hunt. This time, they had to be completely revised and rewritten: twenty-six new articles were added, and fines and other penalties were provided for owners and officers who failed to comply with them. Shipowners were now obliged to feed the men while they were preparing the boats for the ice, to provide them with sanitary sleeping quarters, to carry a doctor on board each vessel, to compensate wounded men and to pay for the funerals of the dead. There were also regulations governing the storage of dynamite and the handling of firearms, while quotas were fixed for the slaughter of adult harp and hooded seals. All in all, the text of the law filled eleven pages of fine print!

With so many restrictions, the seal hunt was to become a much less lucrative enterprise for the merchants, who found themselves jostled by events on all sides. War had broken out in Europe, and the Newfoundlanders were joining the armed forces. Blow for blow, they preferred the battlefields of the Marne and Flanders to those of the icy Front! It became increasingly difficult to recruit competent crews for the sealing vessels. Soon even this ceased being a problem, for England requisitioned all the steel-hulled boats for military service. The *Stephano* was torpedoed by a German submarine on October 8, 1916; the *Florizel* went down in February, 1918; the *Adventure*, the *Nascopie* and the *Beothic*, used to transport troops overseas, were sold at the end of the War to British interests – a more profitable alternative to bringing them all the way back to Newfoundland.

The oil market, too, was shifting: black gold or petroleum, the modern panacea, was quickly gaining ground on seal fat and spermaceti. The London market, which had thus far controlled the world-wide commerce of these products, suddenly found itself facing stiff competition. The wholesalers were slow to understand that it would no longer be profitable to exploit the seal for its fat alone, that it would also be necessary to make use of its fur. By the time they awoke to this fact, Oslo had replaced London as the world's sealing capital.

As a result, the merchants of Water Street gradually lost interest in the seal hunt. They directed their investments instead toward more dynamic sectors of the economy: insurance, maritime transport, commerce, banking, and real estate. Though they continued to send their old "wooden walls" out to the ice each spring, their hearts were no longer in it; they were simply letting the vessels end their days at the Front.

To this end, they were often successful. Between 1918 and 1931, nearly a dozen boats, including the *Nimrod*, the *Diana*, the *Kite*, the *Nord*, the *Erik* and the *Newfoundland* itself (renamed the *Samuel Blandford*) perished in the ice, with a heavy loss of life. Occasionally, a merchant would take it into his head to equip a steel-hulled vessel for the hunt, like the famous *Imogene*, which brought back 55,600 pelts in 1933; but these were mainly passenger ships used in the sealing campaign only during the off-season. There were simply not enough seals left to maintain the old-style sealing fleet.

During the 1930s, the Norwegians began to cross the Atlantic to hunt the herds of the Gulf and the Front. Great scavengers of the world's oceans, they had so completely decimated their own colonies of harp and hooded seals[18] that it was now more profitable for them to undertake the long trans-Atlantic voyage to scrounge for the remnants of the once-bountiful resources of the Northwest Atlantic! They tore into the ice-floes with their deadly little ice-breakers, and their attacks on the herd would almost certainly have signalled the end of the great pagophile nation if the Second World War hadn't intervened to put a temporary halt to the carnage.

Never had *pagophilus groenlandicus* and *cystophora cristata* found themselves so close to extinction as they were in the spring of 1940. The demographic curves were moving steadily downward to the point of no return.

The misfortune of one species proved, however, to be the good fortune of another. While the men were devoting all their energies to destroying one another, the seals enjoyed a period of salutary peace. For five years, the seal hunt virtually ceased: the sealers were in uniform, their boats had been requisitioned by the various warring powers, and the mine- and submarine-infested waters of the North Atlantic discouraged all unnecessary crossings. Only the aborigines and the inshore sealers continued to scour the ice-floes for blubber and furs. In a little over ten years, the herd doubled in size, rising from less than 1,500,000 heads in 1940 to nearly 3,000,000 a decade later.

Once peace had been restored, however, the hunt was resumed on an even greater scale, as if the merchants felt obliged to make up for lost time. The logistics of war so recently tested on the battlefields were now turned against the seals: the time-honoured seal hunt took on the appearance of a campaign of wholesale butchery. Attacked from the sea and the air, the beasts stood no chance. The sealer, who had once had to call upon all his ingenuity to locate the herd and to transport his pelts back to shore through the innumerable snares of the living ice, now found himself reduced to the role of a mere cog in the immense arsenal of machinery and technology launched by industry against the seals.

Ice-breakers, airplanes, helicopters, skidoos, all converged on the ice, guided by radar and other newfangled instruments of electronic navigation. The sophistication of the arsenal made the wooden club still used by the sealers appear anachronistic, cruelly old fashioned. And yet, it remained an undeniably effective weapon, even in this new age of technological innovation. By the hundreds of thousands, the pagophiles were clubbed down: 439,000 in 1951; 334,000 in 1955; 391,000 in 1956; 342,000 in 1963. The following year, the 340,000 harp seals slaughtered on the ice brought the losses suffered by the great herds in the twenty years following the War to more than 5,000,000 head.

And all that was left to the Magdaleners, Newfoundlanders, and Nova Scotians who executed the carnage were bloody hands, a sea increasingly devoid of biological resources, and a few pennies to keep body and soul together. The fortunes were amassed on the other side of the Atlantic, by the owners of the Rieber Tannery in Norway, after their suppliers, the Carino and Karlsen firms (whose interests in Newfoundland and Nova Scotia were looked after by a consortium of Norwegian immigrants and naturalized Canadians) had handsomely padded their own bank accounts.

At this rate, it seemed likely that the harp seals would be extinct by the end of the century. But times were changing, more and more rapidly, in this new era of change and speed!

In 1964, a team of Canadian cinematographers came to the Magdalen Islands to film footage for a television documentary on the seal hunt. This was at the height of the airborne hunts and their film graphically reflected all the insanity of the campaign.

Made theatrically vivid by the stunning backdrop of the living ice, the sad little faces of the "baby" seals, the violent contrasts of the red blood, the black and green oilskins, the yellow airplanes, and the blue sky against the dazzling white ice-pack, punctuated with scenes of brute violence, and documented with alarming statistics, the film resulted in an immediate outcry throughout the entire civilized world.

Thereafter, each spring, legions of film-makers, reporters, photographers, and sightseers invaded the ice, aggravating the controversy. Overnight, the "baby" seal became an international celebrity, appearing on the front pages of newspapers and on the covers of many of the world's leading illustrated magazines.

Today, looking back on all this, though I still think that the campaign on behalf of the "baby" seals was a deliberate low blow to the Magdaleners, who were made to look responsible for a crime of which they were merely the instruments, I am inclined to view all the slander and abuse showered upon the sealers a little more indulgently; for the situation was an urgent one. It was necessary to strike, and strike fast, to save the harp seal. What if the blow was struck in the wrong place? What if it resounded with dubious, demagogic arguments based more upon emotion that reason? What if, through an excess of anthropomorphic zeal, the abolitionists even went so far at times as to equate the "baby" seal with Vietnamese and Biafran infants? In 1971, there remained no more than 1,255,000 seals in the great herds that, two centuries earlier, had numbered ten times that many; and they were still being slaughtered at the rate of more than 200,000 per year!

Seen in retrospect, the crusade to abolish the seal hunt had results that were, at best, paradoxical. Ironically, by saving the seals, it made it possible for the hunt to continue.

Faced with a rising wave of public indignation pressing for action, and unable to ignore the fact that it was in Canadian waters that the genocide of the seals was taking place, the government of Canada acted as if the situation would eventually take care of itself. Granted, there were token gestures of appeasement: the Minister of Fisheries, M. Robichaud, was dispatched to the Front to personally look into the situation. He spent an entire afternoon on the ice, surrounded by the bigwigs of the sealing industry. Needless to say, he saw nothing alarming. Upon his return to Ottawa, he informed Parliament that the slaughter of seals was carried out in a humanitarian manner, and that the seal hunt made a valuable contribution to the economy of the regions in which it was conducted.

Taking these observations at face value, the Liberal government of Canada adhered to all the best principles of liberalism ("Liberalism," a philosopher said, "is freedom: the freedom of a fox set free in a free henyard"): it let things be. Meanwhile, international public opinion remained stubbornly opposed to the hunt, and the Magdaleners found themselves sucked deeper and deeper into the quagmire of sensational journalism.

Skilfully orchestrated by a few leaders who had a great deal to gain from the noble crusade, the controversy gained momentum. Having made headlines in the newspapers and prompted mass demonstrations in the streets of several European capitals, it soon found an attentive ear in a number of embassies. Diplomatic notes of protest were addressed to the government in Ottawa, which was finally forced to act in 1969.

This time, it came down hard. Its intervention in the seal hunt was every bit as rigorous as its past neutrality had been lax. It outlawed the use of airplanes and banished the big commercial sealing vessels from the waters of the Gulf, confining them to the region of the Front, where they were subject to strict quotas. The harp seals were saved.

In an attempt to silence those who claimed that the whitecoats were being skinned alive (though scientists agree that the twitching of the body sometimes observed during the sculping process is a post-mortem reflex caused by the powerful anaerobic metabolism of the animal) the government also passed a law forbidding the use of the gaff. Then it organized a Committee on Seals and Sealing, to which it appointed a number of eminent scientists, as well as representatives of various humanitarian organizations, and instructed it to report back as soon as possible on all aspects of the question.

In the Magdalen Islands, these measures were sufficient to restore the hunt to its old artisanal ways. Except for the helicopters that continued to scour the skies – filled now, not with sealers but with biologists, official observers, wealthy tourists, and an endless string of civil servants and bureaucrats dispatched to the Gulf by the Department of Fisheries – it almost seemed that the clock had been turned back. The inshore sealers exhumed their *canots* and their ropes and took once again to the water, paddling far out to sea in search of the scattered remnants of the great herd. The little twenty-metre trawlers – the largest sealing vessels now permitted in the Gulf – were once again pitted against the ice-pack. Almost overnight, the great marine sealing tradition that had been lost during the era of the aircraft was miraculously revived.

There was something very poignant about the sight of this little flotilla of battered boats leaving the habour of Cap-aux-Meules each year at the beginning of March. Their motors chugging, their sirens and bells ringing out a farewell, they were a stirring reminder to one and all of the inequality of the combat in which they were about to engage: a handful of frail little crafts on their way out to sea, where, for days and weeks on end, they would thread their way amongst giant fortresses of ice capable of crushing them at a moment's notice. Theirs was a message of courage and fortitude, a reminder to the world that, for the Magdaleners, the seal hunt was something more than mindless butchery.

As if to convince sceptics of the perils of the campaign, three of these vessels – the *Lucy Carmen*, the *Sainte-Lucia*, and the *Cygne* – were sunk by the ice, though without loss of life.

As in ancient times, returning vessels announced the success of their campaign from a distance, flying flags, blowing whistles, firing guns, a shot for every hundred pelts in the hold.

Thus, the exploitation of the seals reassumed human dimensions. Kept in check at either end of the scale by nature herself, a balance was quietly re-established between the pagophiles and their predators inside the Gulf. And while the herd began to replenish its severely attenuated stocks, the Magdaleners, who had learned the bitter lesson of industrial waste, took steps to regain control of this resource that history had long ago bequeathed them.

Modest and realistic though they were, their plans amounted to nothing less than a revolution. Rather than selling the raw pelts for a pittance to Norwegian interests, they planned to process them themselves. First, they would build a small factory where the fat would be removed from the pelts; then, they would construct a number of cottage-industry tanneries, and eventually, a network of workshops to manufacture articles from the processed pelts. In this way, the profits from the seal hunt would remain on the Islands, while stimulating employment and creativity. No one, not even the staunchest adversary of the sealing campaign, could find fault with such a project; for what is improvident about a people living off the sensible exploitation of the natural resources that surround it? Man, too, it must be remembered is a part of nature.

As proposed by *l'Association des chasseurs de loups-marins des Iles de la Madeleine* (The Association of Sealers of the Magdalen Islands) this project received the prompt approval of the Committee on Seals and Sealing. Indeed, it was so whole-heartedly supported that, when, in 1972, this Committee recommended a moratorium on the seal hunt for a six-year period (the biological cycle of the seal), it was tacitly understood that the Magdaleners and the aboriginal peoples would be exempt from this measure. And when, in 1974, the then Minister of Fisheries, Mr. Jack Davis, rejected the Committee's recommendations and decided not to institute a moratorium on the hunt, it was made to appear that this decision was motivated by a desire not to interfere with the Magdaleners' modest projects!

As if the magnates of the fur industry, the powerful lobby of sealing interests, and the Norwegian Embassy had no sympathetic ear amongst the entourage of the Minister.

Thereafter, the Magdaleners found themselves living on dreams and promises. It was an education in itself to hear the stream of civil servants who came to "consult" with the sealers: the pledges, the assurances, the endless lip service! The Gulf herd was ours, we were told; we had inalienable historic rights to it, there would never be any going back on that. As soon as its stocks were sufficiently replenished, we would be allowed to send bigger sealing vessels out once again, to ensure our projected factory a constant supply of pelts. But in no case would these boats be allowed to take precedence over the inshore sealers; they would not be permitted to work any patch of seals that looked as if it were heading for land; they would have to wait until the landsmen's quotas were filled before being allowed to go after the seals. For the Canadian government was much more concerned with protecting the interests of the little man than those of big industry, that much must be made very clear!

Not all civil servants were of the same opinion. While some opened the door on the wildest hopes, others came along to slam it shut. They had to be seen to be believed: lecturing an assembly of sealers, brows furrowed, fingers wagging in the air, like schoolteachers reprimanding their pupils. According to them, it was the inshore campaign that had given the seal hunt its bad name around the world; it was the inshore sealers who were responsible for skinning whitecoats alive, using outlawed hunting equipment, and behaving like savages on the ice. It was they whom government helicopters had to rescue from the difficulties in which they were constantly finding themselves. It was they who dressed the sculps so clumsily that they lost up to three-quarters of their market value. And, there and then, they threatened the sealers with outright abolition of the inshore hunt if they did not obey the often ridiculous, sometimes homicidal (I have mentioned the gaff, haven't I?) regulations they were forever concocting.

Now, there is a misery worse than the misery of destitution: it is the misery of being the plaything of technocrats and bureaucrats, of being dragged endlessly through the labyrinth of administrative intrigue. For nearly five years now, the Magdaleners' project for the indigenous development of the sealing industry has remained up in the air. Like a football, it has been passed from one governmental department to another, engaging whole teams of civil servants in investigating, analyzing, studying, surveying, filling endless files with useless paper work – while those they are supposed to be serving are obliged to sit back and wait.

And while the files play hide-and-seek from office to office – lying in the bottom of a desk drawer, stagnating beneath a pile of more urgent paperwork, serving as fuel for some internal dispute between feuding bureaucrats – other civil servants, with other files, find an attentive ear in the Minister's inner circle, and once again the promises fly.

In 1977, it took no less than two dozen civil servants to come to the Magdalen Islands to announce to the sealers that the Gulf of St. Lawrence would be reopened to large sealing vessels. Things were looking up: recent demographic studies indicated that the seal herd had sufficiently recovered its losses to justify a more intensive hunt. Indeed, an increase of quotas had become imperative because of the harp seal's pressure on commercial species of fish. In light of this, the government had decided to reissue permits to large sealing vessels in the Gulf. Of course, these boats would have to be registered in Quebec, and they would have to be manned by Magdalene crews. And at no time would they be allowed to interfere with the inshore sealing campaign.

That spring, everything went well. The herd remained strung out at sea and the inshore catch was minimal – so small, in fact, that the government issued a second permit to the large sealing vessel *Nadine* to profit from the seals the landsmen had been unable to harvest.

The following year, alarmed by the possibility of a "demographic explosion" of harp seals and the unfortunate short-range repercussions this might have on the already unstable fishing industry, the Canadian government decided to issue large sealing vessels second permits in the Gulf. In spite of all the fine promises that the Gulf herd belonged to the Magdaleners, it was handed over to the powerful Karlsen Company of Nova Scotia. And in spite of all the fine promises that the inshore sealing campaign would always take precedence over big industry, the hunt was closed to the landsmen as soon as they reached their quota of 15,000 pelts – while the seals still swarmed all about the archipelago – in order to give industry "a chance"!

This was difficult for the Magdaleners to swallow. They had

been led to believe that limits would never be set on the landsmen's catches as they fluctuated greatly from year to year. Only once every five or six years, does the catch exceed 10,000 seals; in a slow season, it might be as low as a few hundred, even less.

This chance was interpreted by the Magdaleners as a chance for the Karlsen Company to exploit the local sealers, and they weren't far wrong. By way of example, here is how the profits are divided – it's difficult to speak of sharing – on a large sealing vessel owned by Karlsen Shipping Company. First, the owner and his captain take seventy per cent of the gross receipts: the boat's share, they call it, though it would be more accurate to call it the lion's share. The rest is divided equally amongst the crew. But out of their modest earnings, the men must still pay for the costs of the trip: their knives, their oilskins, the food they consume during the two or three weeks of the hunt, the fuel burned by the boat, the gasoline for the skidoos, and another fifty dollars a day for the time it takes to unload the pelts, if they aren't willing to perform this arduous task themselves. Once the pelts are carefully stored in the company's warehouses, all they have to do is fork out the price of a return ticket from Halifax to the Islands.

The local shipowners are less greedy: they take only sixty per cent of the gross receipts and they cover the costs of fueling and unloading the boats. A crewman from the *Nadine* showed me his pay cheque: $3,500; a crewman from the *Martin Karlsen* showed me his: $1,733. And yet, that year, the two vessels had harvested an equal number of seals!

And thus, the wheel had come full circle: scarcely ten years after having been banished from the Gulf for the abuses they had committed there, the big sealing vessels found themselves once again ruling the roost, this time with the blessing of the Canadian government. The inshore sealers had no choice but to crawl back into their holes, taking with them their dreams and their stillborn projects for the indigenous development of the sealing industry. You have to think big in this world to succeed!

There are now six players in the swiling game: the seals, the sealers, industry, the government, the scientists, and the abolitionists. It's like a hockey game in which each team tries to win points at the expense of the others – with this distinction: the players in this particular game can change sides at will. Thus the government, which tries to alleviate the fears of the abolitionists while making empty promises to the inshore sealers of the Magdalen Islands, can suddenly pass the puck to big industry. The biologists tried to pass it to the seals, but without scoring a point. Likewise, the abolitionists, the indomitable friends of the seals, can find it in their hearts, one cold night on the ice with their helicopter inoperative, to slaughter a few of their protégés to keep themselves warm over a blazing fire of seal fat – a fact that was witnessed by an officer of the Fisheries Services in the spring of 1977! Why should it come as a surprise, then, if one day the inshore sealers should decide to strike a deal with the ecologists – provided the latter overcome their prejudices – in order to check big industry and to aid them in their modest project for ecological self-sufficiency?

No, there would be nothing particularly surprising in that. I once heard a spokesman for industry inform a team of civil servants that, if he were not granted a permit for his big sealing vessel, it would be more profitable for him to align himself with the abolitionists. All he was saying, in his own way, was what the magnates of the sealing industry never tire of repeating: the seal hunt will collapse if they are not allowed to hold a monopoly on it.

XVI
Seals and Captivity

In recent years, some seals have had the misfortune of being adopted by their benefactors rather than dying beneath the sealers' clubs.

In the mid-1960s, for example, two whitecoats born near the Magdalen Islands were taken under the protective wing of one of the more ardent adversaries of the seal hunt. Given little-dog names and installed in the ice-cube filled bathtub of their inveterate friend, hundreds of kilometres from their natal ice-pack, the two little beasts were quickly set to work to illustrate the great love that can exist between seals and men, when the former are not denied it, as they are every spring on the ice. A long procession of journalists, photographers and film-makers filed through the home of the celebrated guardian of the phocidaec race to record on paper and film such moving scenes of familial happiness as the survivors of the ice being cuddled and caressed by his charming wife and children. How touching to watch those silky little pups romping on the living room carpet, before a roaring fire in the grate, the object of so much love and affection!

Alas, the metabolism of a seal is not designed for suburban comfort. After a short while, one of the pups began to weaken: its eyes clouded, its moustaches drooped, its hair grew flaccid, it refused to eat, and it spent its days crouching in a corner of its miniature ocean, as if paralyzed. A few days later, it was dead.

This pup must have had a feeble constitution, for its companion, subjected to the same treatment, continued to thrive, so that it soon began to outgrow its rather cramped living quarters. The waves of sight-seers had fallen off and, with no one to record the happiness of this mixed household, it is understandable that its self-appointed saviours began to lose interest in their charge.

Some months later, the great protector of the pagophiles made it discreetly known that he would like to be relieved of his protégé. The young beater was acquired by the Vancouver Zoo, where it died less than a year later.

More recently, a great lady of the screen also became the custodian of a seal.

A Breton fisherman, aware of the famous actress' quasi-maternal affection for amphibians, sent her a common seal he had discovered in his nets and that was of absolutely no use to him. Deeply moved, the forty-year-old beauty christened her protégé *Chouchou* and installed him like a visiting dignitary on one of her estates. The latest in a long line of beaus, darling *Chouchou* led a pampered existence, ensconced in his mistress' swimming pool with its blue chlorinated water, cuddled, coddled, fed the finest fish, with nothing to do all day but perform for the endless amusement of his hostess' guests.

But even the best-intentioned of humans tire of things in the long run, and the great lady soon found herself with other fish to fry. Pressing business called her away. Leaving her seal in the custody of a caretaker, she departed to tend to her affairs. Was this man lacking in scruples, did he dislike animals, was he negligent? Who knows? But when the biologists of Marineland in Antibes accepted delivery of *Chouchou*, shortly after his mistress had abandoned him, he was suffering from extreme malnutrition and was near death. Serums and vitamins had to be administered to the little creature to keep him alive.

Today, recuperated from his ordeal, *Chouchou* entertains visitors at the Antibes Zoo. But let no one suppose he is happy, turning in circles in a tiny pool and diving for coins thrown by the spectators, despite signs prohibiting this practice. Like all seals in captivity, he leads a miserable existence, not unlike that of a man in prison. Despite being looked after by a trained and attentive staff and being fed all he can eat, he will always lack the one element essential to his happiness: the freedom of the open seas.

One of the clearest indications of the poor moral and psychic health of seals in captivity is their refusal to mate – when they live long enough to reach sexual maturity, that is. Of course, there are exceptions to this rule, like the pair of grey seals in the Quebec Aquarium that gave birth to a pup a few years ago; and the odd harbour seal that has reproduced from time to time in one or another of the world's zoos. But these two species are more or less sedentary by nature, even choosing at times to live in fresh water, so the atmosphere of an aquarium may not seem too alien to them. In the absence of ice, grey seals can whelp on rocks, as often happens at Deadman Island.

Harp seals and hoods, on the other hand, have never been known to whelp anywhere but on the ice.

Besides, they never live long in captivity. The harp seals survive a few years; the hoods, a few months. Of course, this doesn't prevent the various animal detention centres from persisting in their attempts to keep these little hobos of the sea alive in small concrete basins. Indeed, they seem to look upon it as a sort of challenge, as if humanity would conceivably benefit from their success in domesticating them to this point. Wellie receives orders for new specimens every year.

I visited the seals one year at Quebec's Aquarium. It was May, before the tourist season, and the seal's quarters were temporarily closed to the public. I believe a reservoir was being built for the harp seals that had recently arrived from the Magdalen Islands, for I saw five or six beaters and one hooded seal pup splashing about in a little basin no more than a quarter full of water.

I remembered this pup, for he had been my travelling companion on the boat from Cap-aux-Meules to Montreal. A magnificent creature, he weighed nearly a hundred pounds, with a marvellous slate-blue coat and big moustaches protruding from his plump nostrils. Short-tempered and irritable, he had roared and bared his fangs the moment anyone had approached his cage. He had refused to eat during the entire voyage.

And now, there he was, lodged with a half-dozen harp seals in a pool containing less than four feet of water. They were all the same age, but the harps were members of a species with which his own had never associated. And while the beaters gamboled about, full of exuberance and apparently happy, he lay obliquely in the water a few inches from the cold water intake, immobile and taciturn. Whenever a beater mistook him for one of its fellows and tried to engage him in play, he greeted it with a display of great animosity, then returned at once to his station by the cold water pipe. Every five or six minutes, he would dart like a torpedo to the far end of the tank, obliging the beaters to get out of his way, then rush back to take up his position near the cold water intake, only his snout sticking out of the water, like a little old man shivering beside a hot stove with just the tip of his nose peeping out of his muffler.

Watching him, I thought of all his brothers and sisters, born like him beneath the spring moon and labouring now in the polar ocean, swimming with all their young strength for the shores of Greenland. And I couldn't help thinking that he would have been more fortunate if I had encountered him somewhere on the ice and crushed his skull with my club. He certainly would have suffered less.

The young hood is dead today. The summer heat proved too much for him. He died, like dozens, hundreds, thousands of others, who have expired in one or another of the world's zoos, amusing the crowds. A sad end. And yet it is infinitely preferable to that of their fellows who end up in research laboratories or the circus ring.

In the confines of a zoo, at least, the seals die peacefully, in more or less comfortable surroundings, the object of a thousand little attentions. Of course, their health is never good: most suffer from gastritis or perforations of the intestine, as a result of ingesting incredible numbers of small coins thrown to them by the public. And their skin is always covered with rashes and sores, from rubbing against the rough concrete walls of their basins. These lesions do not easily heal and often become infected in the excrement-polluted water. The chlorine added to the water to disinfect it merely causes other problems, such as the eye ailments from which most seals in captivity suffer. This is to say nothing of the great variety of psychological and emotional disturbances to which they are subject, but which are always more difficult to detect. Phocidaec psychoanalysis is still in its infancy stages.

In circuses and research laboratories, aside from suffering from all the same ailments, seals are often subjected to treatments that are less than humanitarian. Circus seals are whipped on the snout until they learn to spin balls on their noses, at least that is how they were trained at the turn of the century; I don't know whether methods have improved since. And laboratory seals are used like guinea pigs in the noble cause of scientific research.

It is there, in the laboratory, that they suffer the most.

Eight years ago, a team of scientists visited the Magdalen Islands to conduct an experiment on seals, which involved confining a number of whitecoats in a metal pen at the old biological station of

Gros-Cap. There they were left for days on end, mewing and whining, without food. I happened to be researching an article on the seal hunt at the time, and I questioned the officer of the Fisheries Service responsible for the enforcement of sealing regulations on the Islands about the treatment to which these creatures were being subjected. Unable to conceal his embarrassment, he replied that it was a top-secret scientific experiment. It was cruel, he knew it, but what could you do? Yes, even he thought so, a man who in his twelve seasons as supervisor of the hunt had witnessed all sorts of behaviour, and who, moreover, considered the seals to be a pest, harmful to the fishing industry. There was something decidedly inhuman about subjecting these little creatures to such prolonged and deliberate suffering. But he assured me that the goal of the experiment was a laudable one. If, as a result of their ordeal, the famished whitecoats excreted a certain substance, x, y or z, it might be possible to save the life of a wealthy patient in Florida who was dying of cancer – or chronic heart disease, I no longer remember.

The pups died without secreting the miracle drug. And the wealthy patient, I am told, followed them shortly to the other world.

This is far from an isolated example of the cruel treatment of seals at the hands of the scientific community. When the supertanker *Argo Merchant* ran aground a number of years ago, spilling 7.5 million gallons of crude oil into the Atlantic and threatening to pollute the habitat of the colony of grey seals on Nantucket Island, a learned Canadian biologist was curious to know to what extent the seals might be able to adapt themselves to the pestilential tide. That summer, high in the Canadian Arctic, far from the indiscreet and critical eyes of those who know nothing about pure science, he had a large tank filled with crude oil and placed three harp seals in it. Then he stood back with a stop-watch and calmly watched them die – a process which took only a few hours.

Only later was this biologist brought to his senses, when, planning to repeat the experiment with a new sampling of "guinea pigs," to prove beyond the shadow of a doubt that seals cannot survive in crude oil, he was dissuaded by a colleague who pointed out that, in fact, seals are excellent swimmers and would flee the oil slick rather than trying to adapt to it.

Since the seal became a celebrity on the international scene, it has been the object of boundless interest on the part of the scientific community. Previously, you could count on the fingers of one hand the number of scientific papers published on the subject in a decade; today, the fingers of both hands would not be enough to ennumerate all those that appear in a single year. Every species of seal, every aspect of the seal – from the retina to the inner wall of the duodenum – have become a source of endless fascination to the scholars and the researchers. More than five thousand pagophiles are captured each year, dead or alive, for scientific purposes – a third of the quota allotted the inshore sealers of the Magdalen Islands!

This sudden interest in the pagophiles is not surprising. Scientists are men, like any others, and despite their professed disinterest, is it not possible that they are motivated to some degree by the vague hope that some of the little animal's great fame will rub off on their work? There is much envy and rivalry amongst faculties, schools, and research institutes, as is seen in the verbal exchanges whose thinly disguised acerbity often exceeds the bounds of healthy competition. (One of the world's great authorities on the seal, the author of an exhaustive 1,500-page bibliography on the subject, is nicknamed "Super-Seal" by his colleagues.) On the other hand, there is a strong sense of solidarity amongst the members of this erudite world, as witness the explosive laughter that invariably greets the now somewhat overworked question, "Who are the more numerous, the men who hunt the seals or the men who study them?"

Of the five thousand seals captured each year for research, most are slaughtered on the ice; and I have nothing to say about them, for they are put to death in a humanitarian way by experienced sealers. It is those captured alive whose fate seems less enviable to me.

Pity the poor seal that finds himself alive on a dissecting table of a university laboratory! Since there are too few specimens to go around, it must serve concurrently for a number of experiments: when its eyes have been removed by one scientist so they can be scrutinized beneath a microscope, it is passed on to another scientist, who proceeds to make studies of its fur (a process which involves removing samples of the animal's skin at fixed intervals) or to conduct experiments on its auditory system, its digestive tract, its kidneys, its spleen, its pancreas . . .

In researching this book, I came upon a collection of scientific papers[19] given at a Symposium on the Biology of the Seal, at Guelph University, the Mecca of seal studies, in the summer of 1972. They confirmed my suspicion that the sealers are not the only ones guilty of cruelty to the seals.

One of the first problems encountered by these scientists in their audacious experiments (such as the conducting of encephalographic studies by placing electrodes in the animals' brains and attaching them to transmitters strapped to their heads), was in anaesthetizing their subjects. While it stands to reason that a creature whose nostrils habitually remain closed in water will not open them when placed under a mask filled with ether or chloroform, researchers first attempted this very approach. Now, a seal can hold its breath for up to half an hour, so it's not hard to imagine who began to nod off first! Then, drugs were administered to the animals, with no idea of their possible effects, for the metabolism of pinnipeds is very different from that of terrestrial mammals. After repeated failures, documented in graphic detail in these reports, an effective anaesthetic was finally discovered.

Unfortunately for the seals, the dissecting rooms and laboratories of the average research centre are not hospital operating rooms, and the biologists who man them are not surgeons. To insert the aforementioned electrodes in the seals' brains, it was necessary to remove the top of their skulls with a special saw. In one reported instance, a biologist operating on a female grey seal inadvertently pushed the instrument too far and injured her cortex. The following day, her head was grossly swollen and she appeared to be paralyzed. A dose of chloromycetin was then administered to prevent infection and the surgery continued – without anaesthetic this time, since the biologist considered it unlikely that her suffering could be intensified!

There it is, in black and white: the animal began to tremble, she was shaken with a series of spasmodic convulsions. Rather than administer a sedative, the biologist profited from the occasion by taking readings on his encephalograph of her brainwaves. So much for the advancement of human knowledge! Three days later, the creature was relieved of her sufferings by death.

The other seals used in the same experiment were more fortunate and provided the researchers with valuable information about their senses of hearing and smell, aggressiveness, sleeping habits, and tolerance to electric shocks.

When the experiment was completed, the seals were returned to the operating table to have the electrodes removed, along with the transmitters that projected from their foreheads like the horns of unicorns. Alas, in the euphoria of the moment, the operation was performed under less than hygienic conditions. Three of the seals developed abscesses on the brain and died.

Those seals who survived the ordeal were not set free, however. They were destined for further experimentation. As soon as they had regained consciousness, they joined twenty-four other harp seals whose eyes had already been removed for study in a group experiment designed to determine the tolerance of pinnipeds to synthetic food. How interesting to learn that the endurance record for this diet is held by a young common seal, who survived a full four months before developing symptoms of colic. Found to be suffering from an ulcer, it was given antibiotics and died, just like all the others.

The cruel and barbarous treatment of seals in the name of scientific research is such a widespread phenomenon that it was actually the subject of a paper given at the Symposium at Guelph University. The authors reveal that laboratory seals suffer from a bewildering array of ailments: bone diseases, the result of poor calcification; sight disorders, ranging from inflammation of the conjunctiva to ulceration of the cornea (treated by smearing the eyeball with medication and stitching up the eyelid!); dental decay, cutaneous complaints, poor digestion, hepatitis, inflammation of the pancreas, ovarian tumours. The description of each ailment is accompanied by a list of the precautions to be taken to avoid it. Thus, I learned that to prevent cataracts or kidney stones in a seal, you must simply refrain from feeding it whipped cream!

In the USA, experiments involving seals take on a more practical character. It would seem that researchers south of the border are less enamoured of science for science's sake than are their Canadian counterparts. They are more interested in the uses to which the beast can be put, the ways in which it can benefit mankind.

In California, for example, sea-lions are trained to retrieve objects from the bottom of the sea, at depths beyond the reach of human divers. First, they are taught obedience, like a dog, through the process of reward and punishment (if you do as you're told, you eat; if you don't, you go hungry), the chain of Pavlovian conditioned reflexes unfolding a link at a time. But these sea-lions are not as docile as their terrestrial confreres, appearances notwithstanding; ungrateful beasts that they are, they have actually been known on occasion to bite the hand that feeds them! In such instances, it becomes necessary to administer a dose of librium to curb the animal's viciousness. Alas, this proves only partially effective, for the drug causes the creature to be lazy and unresponsive.

Those subjects who successfully complete their preliminary training are ready for the harness – a cumbersome, ungainly prosthesis strapped to the body of the beast and terminating in a muzzle equipped with large pincers or forceps controlled by the movements of the jaws. Frogmen teach the sea-lions the use of these pincers, beginning with tiny objects in shallow water. When they have mastered the technique, they proceed to retrieve larger objects at greater depths, being kept at all times on a leash. For it would be a shame if these creatures gave their human friends the slip after all the time and effort expended in teaching them their art. At last report, the most talented sea-lions could dive to a depth of several dozen fathoms and return with objects a foot in diameter and weighing up to a ton.

Who said the talents of these beasts were fit only for the circus ring?

I don't wish anyone to conclude from these derisive remarks that I reject the work of researchers en masse. On the contrary, I have the deepest respect for those who have studied the population dynamics of the seal herds, and conducted experiments on the feeding habits, the ecology, the ethology, and the sensory perceptions of the pagophiles; and I raise my tuque to all those biologists who accompany the sealers out to the ice each spring to tag, ring, and count the seals, so the annual seal hunt will never again be allowed to push the species to the edge of extinction.

Some experiments even have their attractive side, like the one conducted for four years on the Magdalen Islands on the underwater communication of the pagophiles during the mating season. Ultra-sensitive hydrophones were placed beneath the ice, at various depths, to record on tape and oscilloscope the multitude of sound waves passing to and fro in the "world of silence." Interspersed with the clicking sounds of opening and closing shells, scientists were able to discern the "conversation" of the seals.

So Thaddée Snault and Abbé Allain had not been dreaming when they had heard the seals talking while en route to their new homeland: these animals do communicate in the wild (though apparently not in captivity), exchanging "words" and patterns of "words" that would appear to be intelligible to them. At least a dozen distinctive sounds, in a variety of combinations, were identified by the researchers, suggesting a rather extensive vocabulary, perhaps even the ability to communicate in sentences.

Close study revealed a certain rigorous protocol in these verbal exchanges. The seal who is about to speak always begins his discourse with a "click," a short, staccato sound that indicates his desire to be heard. Then he launches into a complex cantata of trills, chirps, and short and long "beeps," like the signals in our own Morse code, interspersed with grunts, squeaks and a great variety of bird-like warbles and coos. When he has finished speaking, he makes another "click" to signal the end of his discourse.

There is a moment of silence and then another "click" is heard, followed by another tirade. A second seal is answering him. Never do two seals speak at the same time.

This discovery raises a great number of intriguing questions: who are the harp seals and all those other seals that inhabit the oceans and seas of this planet, from the poles to the equator? Will the key to their mysterious language ever be discovered? Will the depths of their history ever be probed? Why, several million years ago, did they part company with the other mammals to return to the sea?

Grateful to the researchers for their ground-breaking work in these fields, it is only human, I suppose, to remain silent about the cruelty and barbarity of some of their methods, provided the results of their experiments are deemed worthwhile. The reason I deplore the experiment in which the eyes of twenty-four harp seals were removed to allow the functioning of their retinas to be examined beneath a microscope, is simply that the poor beasts were not immediately put to death. For I must admit that I, too, was interested to learn that the photo-receptive cells of a seal's eyes are structured in such a way as to allow it to see as clearly in the dark depths of the ocean as in the transparent waters of its surface, as well as to distinguish objects in murky water, thanks to a sophisticated mechanism that filters ocular information directly to the brain.

No, I am not casting stones at scientists and researchers. But I am a part-time sealer, and as such I am irritated by the myth of cruelty levelled against me and my fellow sealers by misinformed public opinion. Cruelty is a strange thing: is the individual who raises a hue and cry about the so-called massacre of "baby" seals and who continues to heat his house with oil and to fill his automobile tank with gasoline – thereby encouraging, if only reluctantly, the wreck of super-tankers and the triumph of pollution – not every bit as "cruel" as the inhabitant of the frozen seaboard who kills a few seals each year to feed his family?

If Newfoundland goes ahead with its plans to exploit its undersea oil deposits along the coast of Labrador, what effect would a major leak have upon the migratory patterns, and even perhaps the survival, of the Northwest Atlantic colonies of harp and hooded seals? If such a catastrophe were to occur – and they do occur, even in ice-free waters – captivity would probably be the safest solution for the remnants of the great seal herd.

Epilogue

Grand Jean parked his pick-up truck in front of the Fishermen's Cooperative shed. It was an April morning, with still no sign of spring in the air. The wind blew in violent gusts off the water, lifting the fine, granular snow that had fallen the previous evening in thick, swirling clouds, and then sped howling amongst the harbour sheds. Several precocious seagulls stood shivering on their spindly legs down at the end of the jetty.

Despite the early hour and the intense cold, there was a great deal of activity on the wharf at Cap-aux-Meules. The Karlsen Company's little cargo-boat, *Arctic Sealer*, had come into port during the night, and a number of idlers were pacing to and fro at its side, looking more like bears than humans in their heavy winter clothing, stamping their feet to keep the blood circulating.

The arrival of this vessel had aroused a great deal of interest, not only because it was the first boat that spring to break through the ice that had held the Magdalen Islands in its grip since the beginning of February, but also because it was scheduled to depart again that evening with the harvest of the local seal hunt, a fortune which the Magdaleners would never see again and for which they would be paid a pittance.

But there was no talk of that this morning. The men spoke, rather, of the temperature and the daring of the *Arctic Sealer* in threading its way through the ice-pack that filled the bay as far out as the eye could see. They praised the skill of its captain, the solidity of its hull. Then, one thing leading to another, the talk turned to sealing – the adventures and misadventures on the ice, the catches of the various teams – and there was wild speculation about the prices the pelts would fetch that year, though these would not be known until later, when the sculps had been processed and were already on their way to the great Rieber Tannery in Norway.

The current sealing campaign had been one of the best in recent memory. The winds had piled the whelping ice up against the archipelago for nearly a week, allowing the inshore sealers to harvest a prodigious number of whitecoats. The bigger vessels, on the other hand, had reaped nothing but headlines in the local newspapers: one had been crushed between two giant ice-floes; another had been rescued *in extremis* by an ice-breaker, limping back into port with a gaping hole in its prow, just above the water line, big enough for a man to pass through.

Grand Jean stood listening to these conversations, saying little, in deference to these men most of whom were his elders.

"We've got to be fools to let all that go for a song," he said finally. "After all the work we've done. There oughta be a law to prevent a single pelt from leaving the Islands before it's been processed."

"You saw what happened to Danker with his processing plant at Grande-Entrée," someone replied. "The spring he opened up for business, Karlsen raised the price of pelts to $35, though he'd never paid more than $8 or $9 before. At the end of the first year, the plant was bankrupt. It continued in the red for a couple more springs before it was forced to close its doors. It didn't do Danker any good that his old man was a millionaire in New York – he couldn't stand up to Karlsen!"

"And you're the ones who paid for it!" retorted Grand Jean.

"That's right," agreed one of the older men somewhat sheepishly. "For six or seven years after Danker closed up shop, Karlsen didn't pay more than $7 a pelt. He made a killing!"

"That's why there should be a law," repeated Grand Jean. "To stop them from robbing us blind." He tapped two sturdy young lads on the shoulder and glanced toward his half-ton truck. Without a word, the three began to unload the pelts, piling them fat to fat, hair to hair, so as not to spoil the furs.

The old sealers had approached and were examining the catch with expert eyes, manipulating the frozen hides with a cautious tenderness, gauging the thickness of the fat at a mere glance, and sticking their fingers in the flipper holes to measure their width.

"A fine lot of pelts you've got here, Grand Jean."

"They're okay," he said, without pausing in his work. "We spoiled a few at the beginning, but we soon got the knack of it. We were going too fast, that's all."

"That's it! Every young sealer thinks he can sculp a seal in thirty-five seconds flat, just like the old guys who've been at it for years. More often than not his pelts are worthless. Either the muzzle's gone or the tail is cut off or the flipper holes are as big as your head!"

"You could get away with that in the old days," said another. "All they were interested in then was the fat. They sold the sculps by the pound! But today it's the fur that brings in the profits, and if you don't have the muzzle with all the whiskers on it or the wee bit of tail they use to make key-holders, your pelt is worth next to nothing once the fat's been scraped off. Fifty cents is all you'll get for it."

There was a brief silence, time enough to allow each man to formulate the reply that was on the tip of his tongue but that he had the dignity to keep to himself: *it was unjust*! Not one of these old sealers was ignorant of the fact that the fat itself was worth at least five times that much. For several years, seal oil had been selling for up to thirty cents a pound, and a man would have to be a clumsy oaf not to save at least fifteen pounds of blubber from each pelt he sculped. Even those who worked with bread knives and paring knives got that much!

Needless to say, the more conscientious sealers would have been glad to be rid of those hare-brained creatures who ruined every pelt they laid their hands on. It was that minority of would-be sealers who gave the hunt a bad press throughout the world. It was they who compromised the reputation of the Magdalen Islands in the eyes of the big furriers and who caused Karlsen himself to tear his hair out each time he received a shipment of pelts from the archipelago. Not that he lost money on them – he ran little risk of that with the meagre half-dollar he paid for a poorly dressed pelt – but it prevented him from making even greater profits, from selling them at top prices.

"It's the oil that brings in the profits," said Grand Jean.

He wasn't too sure of this, but it seemed only logical, in light of all the fuss being made about the so-called oil crisis. From his perspective, all the by-products of the seal – meat, fat, fur – had essentially the same value, since each in its way allowed man to provide against what he, Grand Jean, saw as his greatest enemy: the cold. If he conceded the oil a slight superiority, it was perhaps because it was intimately associated, in the back of his mind, with a severe thrashing he had received as a child from his father.

He was only a young lad then. In those days, some sealers still melted the blubber themselves in big open vats, pouring the oil into barrels and selling it to the merchants. One day, while playing in the shed where his father kept his fishing gear, Grand Jean inadvertently upset one of these barrels. On his way home, he thought of shifting the blame for his misdeed onto a cat or a big rat, but the powerful odour of seal fat emanating from his clothing left no doubt as to who was the guilty party. The pain he had carried around in his backside for the next week or so taught him once and for all that seal oil is a precious commodity.

"The fat is all they were interested in in the old days," repeated the man who had been speaking earlier. "They didn't give a damn about the rest. But to get back to what I was saying: when our pelts arrive at the plant in Nova Scotia, they're all thrown together, Grand Jean, and when they're processed, there's an expert who examines them and grades them.

"If there's so much as a little blue mark on the hide where it's been grazed by the sealer's knife, right away it drops to the second class. If there's more than one mark or the flipper holes are too big or if the muzzle or the wee bit of tail is missing, it's a third class pelt – fifty cents!

"Then the guy counts them: so many first-class, so many second, so many third. He works out the total in dollars and cents, then divides by the number of hides, and gets the price that'll be offered us for sealskins that year. With a system like that, everything thrown together, the good hunters lose out because of the bad. Which doesn't encourage anyone to do a good job."

There was another moment of silence, then someone added: "In any case, that's a fine lot of pelts you've got there, Grand Jean. No one can claim you bring prices down."

Grand Jean felt a moment of pride at hearing these words, not because they flattered his ego but because they represented the approval of his elders in a field in which he had his heart set on success. For him, sealing was one of the few occupations left in the world in which a man could pit himself against nature; and he had long ago made up his mind that that was the life he wanted to live, even if the resigned looks and defeatist attitudes of his elders often left him with the feeling that he was probably betting on a dream. The trend these days was towards more stable, monotonous, meaningless jobs that offered economic security, the highest possible wages for the least possible work. Adventure, risk, the very salt of life, seemed to have vanished from human ambition. And he instinctively resisted, with his whole being, what most people persisted in calling "progress."

All the same, the disenchantment of the old men was not to be dismissed. For some twenty years now, they had watched this "progress" at work in the once-fertile waters of the Gulf: giant seiners sucking up the herring by the millions of tons; factory-trawlers cleaning out the cod, depleting the haddock, decimating the redfish and halibut. And the large industrial sealing fleets of Newfoundland and Norway indiscriminately slaughtering the seals . . .

And while the Gulf was being exhausted of its formidable resources before their very eyes, they persisted in attempting to make a living from it – though their efforts brought ever-diminishing rewards – by clinging to the artisanal traditions of their ancestors, the only techniques they considered fair in the harsh game of life and death in which they were engaged. Now, the apparently inexhaustible stocks they had once been able to reach with only a few strokes of the oars were depleted, stripped clean. And what were they being offered in return? More progress!

No wonder they were discouraged. Some already looked upon their time-honoured trades as anachronistic occupations doomed to disappear under the tyranny of machines; others were trying to dissuade their sons from following in their footsteps; and everyone faced the future with that sense of crippling fatalism that has marked the mentality of Acadia's children since the Expulsion.

But Grand Jean was still animated by the optimism of youth. All his thoughts and feelings were motivated by a sense of nostalgia for the past. More than anything else, he dreamed of being like those hollow-cheeked men of old, their faces creased by the wind and the salt sea air, their eyes gentle and unclouded, sitting at the helms of their little boats and peacefully riding the waves, happy with their simple, productive lives that were wed to the rhythm of the seasons.

He tossed the last of the pelts from the truck, leaped nimbly to the ground and began to clean his hands in the snow.

"So, where's that guy with the little black book?" he inquired good-humouredly of no one in particular.

"At this hour, he must be still in bed."

"No, he's probably at mass. A man with so many sins on his conscience . . ."

"Unless he's in bed with the priest!"

Now that their work was done, these simple, rough men were in high spirits, like schoolchildren let out for the holidays. They jested, a little facetiously perhaps, though with no real ill-will towards the important personage who was the butt of their joke. They were simply enjoying themselves, doing their best to forget the cold.

Their jesting was cut short, however, by the arrival of a large, shiny automobile that pulled in beside Grand Jean's truck. A stout, ruddy-faced man got out. He was well dressed and carried himself with the assurance of one who knows he has things under control.

"There's your man," said the sealer standing next to Grand Jean, nudging him discreetly with his elbow. All the men greeted the newcomer a little awkwardly.

The Karlsen Shipping Company's agent returned their greeting with a smile and a slight nod of the head. He cast a quick glance over the pelts, then turned to Grand Jean and said: "Are these yours? You weren't in any hurry to bring them in, were you?" His tone was amiable, with perhaps just a trace of reproach in his words.

"They were buried in the snow. They're all right."

"How many have you got?"

"A hundred."

A strange glint came into the buyer's eyes, as if for a moment his pupils had been transformed into dollar signs. He glanced appraisingly at the pile, then turned back to Grand Jean:

"The clerk'll be here any moment. He'll settle with you."

And, without another word, he stepped back into his automobile.

"How much are you offering?" asked Grand Jean, loud enough to be heard over the roar of the eight-cylinder engine. The coolness of this man piqued him even more than the temperature.

"That hasn't been decided yet," replied the agent, opening his car door just a crack. "We won't know for another month. You'll be given a deposit of seven dollars per pelt. If you're owed more later, you'll get a bonus. It all depends on the market."

"That's not a hell of a lot for all the work we've done," replied Grand Jean.

"The price is the same for everyone," said the buyer. "There's nothing I can do about it. We still have a surplus of beaters from two years ago. With seal fur being boycotted in Europe, anything can happen."

The men waited until the merchant's car had pulled away before speaking.

"You better believe there's nothing he can do about it! He gets a commission of $1, maybe $2, on each pelt, and he scarcely has to lift a finger. What are our problems to him? There were 12,000 pelts harvested this year – that's up to $24,000 in his pocket."

"That's no chicken feed."

"He won't go in the red, you can bet on that!"

"And we think we're lucky if we make $10,000 a year, with the seals and the fish and the unemployment benefits all thrown in."

"And he hardly has to lift a finger. A couple of telephone calls and the goose falls on his plate, cooked and ready to eat!"

Grand Jean listened to the recriminating remarks of the other men with a vague smile on his face, as if they scarcely concerned him.

"All you have to do is refuse to sell him your pelts," he said finally. "That way he can't get rich at your expense."

"That's easy to say, Grand Jean," someone replied. "But if we don't sell the pelts, how will we equip ourselves for the fishing season? We'd just go deeper in debt."

"Well, *we* aren't selling any more than we have to. A hundred pelts, that's all Karlsen's getting from us. The five guys on our team brought back 250 pelts. The other 150, we're holding on to."

"And what are you going to do with them, tell me that? What in God's name are you going to do with them?"

"We're going to do like they did in the old days," replied Grand Jean. "We'll wait for the first sunny days, then we'll scrape the fat from the hides, using our sculping knives. Working as a team, it shouldn't take us more than a day or two. Then we'll salt the hides and roll them up and send them by the first boat to the tannery in

Quebec City. That's why we're selling this batch now, to cover the cost of tanning the others. Then, we'll see. Each man will get his own hides back and he can do what he wants with them. Maybe we'll sell them to tourists. Maybe we'll make vests and tuques and mittens of them. Either way, the hides'll bring us more than if we sell them now. There're people on the mainland ready to pay fifty, even sixty-five, bucks for a nice sealskin. That's more than Karlsen offers. And we can sell the oil, too."

The clerk arrived with the punctuality of a clock, at the very moment the siren at the fish plant rang to call the employees to work. Removing an enormous ring of keys from his pocket, he entered the shed, where he remained for a moment, then stepped back outside, armed with a small black notebook and a pencil. He knelt almost religiously before the stack of pelts and began to count them methodically, from top to bottom, from bottom to top, making a small mark in his notebook for each pelt. Then he stood up and totalled the column, checked it, rechecked it, and finally announced triumphantly:

"One hundred!"

"Exactly," said Grand Jean.

"That will be $700. Come to the office this afternoon and your cheque will be ready."

When Grand Jean returned to the quay that afternoon, the pelts were being loaded on the *Arctic Sealer*. Several men armed with shovels were digging them out of the heap of snow behind the shed where they had lain buried, with no other protection from theft than the proverbial honesty of the Magdaleners. Meanwhile, other men were stacking them on wooden platforms, which a noisy little fork-lift truck was carrying at a dizzying speed between the shed and the vessel, sliding its long steel teeth beneath them at one end, lifting them up, then setting them down roughly at the other end, filling the air with an incredible din. On the wharf, they were picked up by a derrick, swung in a graceful arc through the air – like someone waving a handkerchief, thought Grand Jean – and vanished into the gaping hold.

Watching this spectacle, Grand Jean was reminded suddenly – oh, just for a split second, as sometimes occurs in moments of vertigo – of a scene he had witnessed as a child, some twenty years earlier, when he had accompanied one of his uncles one April day with a shipment of pelts to the Havre-Aubert quay.

How times had changed! Only twenty years ago, the people of the Magdalen Islands had still heated their houses with coal and burned oil in their lamps!

In his mind's eye, he saw the steamer *Terre de Fundy* bobbing gently at its moorings and the team of work horses pulling the heavy sleds loaded with pelts and barrels of oil up to the anchored vessel. The memory was so vivid that he could almost hear the squeaking of the runners on the ice, the ringing of the sleigh bells and the harness chains, and the heavy, rhythmical breathing of the horses.

What poetry there was in the past! he thought, while his mind roamed at will over the countryside. Suddenly, amongst the countless images flooding his brain, there emerged the figure of Ti-Médée Arseneau, captain of the *Terre de Fundy*, a legend in his own time, all sinews and nerves, who wore a hook where his left hand had been.

Since the *Terre de Fundy* carried coal between Pictou, Nova Scotia, and the Magdalen Islands, Ti-Médée was always black from head to toe, as he scrupulously avoided all contact with soap and water. "Wash?" he was often heard to say, his voice filled with indignation. "Wash for who? Wash for what? At Pictou, we shovel the coal into the hold and the bunkers. When we're at sea, we shovel it into the ship's furnace. And when we get here, we shovel it off again. So what good would it do me to wash?"

Only once a year was this heretic induced to take a bath. That was during his annual voyage transporting seal pelts to Newfoundland. Then, in addition to the pungent and slightly sulphurous odor of anthracite that normally emanated from him, he also reeked of rancid seal blubber, and each year his nauseated crew found itself on the verge of mutiny.

One night, his men would take action. They would steal his filthy clothing and set fire to it, then seize their skipper and immerse him in a tub of hot water. They would scrub him roughly from head to toe, using a stiff brush and black soap, while he struggled and protested like a devil in holy water, swearing by all the saints on the calendar to set every man ashore at the next port. But, by the time the boat reached its destination, he would have regained his normal charcoal hue and acquired a brighter view of things. All his threats would be forgotten, and relations between him and his sailors would remain cordial until the following year.

One day, while the *Terre de Fundy* was being loaded with pelts, Captain Arseneau's wife, an elegant, somewhat affected lady from Quebec who seemed the last person in the world who would fancy Ti-Médée, came onto the wharf to introduce her husband to one of her female cousins who was on a visit to the Islands. Ti-Médée had spent the entire morning repairing his windlass, and his forearms were smeared with thick black grease right to the elbows. When his wife and her guest reached the gang-plank, he was passing with a huge stack of pelts on his shoulder – Ti-Médée was not one to sit back with his arms folded while his crew did all the work – only the top of his coal-smeared head visible above the rancid blubber.

Taking him for a stoker, his wife hailed him and asked to be taken to Captain Arseneau. He turned brusquely to her and retorted: "And who the hell do you think I am, woman? Yes, it's me, your husband! Idiot!"

Though most women deliberately plaster their faces and anoint their bodies with a variety of products containing a derivative of seal oil, the majority of them have never been very receptive to the olfactory charms of the real thing.

"Ah, Ti-Médée!" thought Grand Jean, as the roar of the forklift brought him abruptly to his senses. "How many men like you are there left, men at one with themselves and the sea?"

Stack by stack, the pelts vanished into the hold of the *Arctic Sealer*. *Au-revoir*, they seemed to say, *adieu*. It was the end of the line, the end of the adventure for another year. And each time another stack disappeared, Grand Jean was reminded that this was the last time they'd be seen on the Islands; and this saddened him, for it meant that the Magdaleners, chronically afflicted with seasonal unemployment, were being deprived of so many possibilities of creative work. All those lovely sealskin articles they saw on occasion in the big city stores, sporting price tags that put them forever beyond their reach, could have been made right there on the Islands!

"Ah well," thought Grand Jean, "that can change, too. Things have changed, they can change again." And he stepped out of his truck to go and pick up his team's pay cheque.

    Pointe-Basse
    May 1977 – January 1979

Notes

1. Etymologically, "friends of the ice." The scientific name for the harp seal is *pagophilus groenlandicus*.
2. There are thirty-three varieties of seal and other marine pinnipeds in the world, of which four, including the grey seal, still frequent the waters off the Magdalen Islands.
3. *Débauche*, or debauchery, is a term used by Acadians to designate a non-working day. (tr.)
4. A sealer's term designating a small, sharp protuberance, a few centimetres in height, on the surface of a smooth sheet of ice.
5. The use of the word *tanner* in this instance has a double significance: it indicates, on the one hand, that the skins of these animals are good only for tanning; and, on the other hand, that they are not of much value, *tanner* being a popular term for a sixpence.
6. A flat beer native to the Magdalen Islands, brewed with water, sugar, and yeast. (tr.)
7. A rather stiff alcohol, of the moonshine variety, distilled from a preparation similar to that of *bagosse*. (tr.)
8. Louis Capet was the civil name of Louis XVI. (tr.)
9. Family names often being identical in Acadian communities, an individual is commonly identified by affixing his father's Christian name to his own. Thus: Thaddée *à* Charles, or Thaddée, *son of* Charles. (tr.)
10. "*Norvagium littus maximos ac grandes pisces elephantis habet qui morsi seu rosmari vocantur. Fortisan ob asperitate mordendi sic appellati, quia sei quem hominem in maris littore viderint apprehendereque poterint, in eum celerime insiliunt, ac dente lacerant et in momento interimunt.*"
11. For example: the grampus feeds on the seal, which feeds on the cod, which feeds on the herring, which feeds on the plankton. But to what extent does the dead grampus lying on the ocean floor provide nourishment for a whole variety of organisms which, devouring each other in turn, bring the ecological cycle full circle by feeding the herring, the cod, the seal, and the grampus?
12. Three young people from the Magdalen Islands, accused the hunting and fishing columnist Serge Deyglun, the director of one of these films, of having gotten them drunk and paid them to skin whitecoats alive on camera.
13. Seal fat has the highest degree of saponification of any product known to man.
14. "*Partie d'iceulx ouaiseaulx sont grans comme ouays, noirs et blancs, et ont le bec comme ung corbin; ilz sont tousiours en la mer, sans jamais pouoir voller en*

l'air pour se qu'ilz ont petites aesles, comme la moitié d'unne; de quoy ils vollent aussi fort dedans la mer, commes les aultres ouaiseaulx font en l'air; et sont iceulx ouaiseaulx si gras que c'est une chose merueilleuse."

15. *"dont chaincun de noz nauires en sallerent quatre ou cing barils, sans ce que nous en peumes mangiers de froys."*

16. Published in Quebec in 1969, under the title *Un Pied d'ancre*, by the author's grandson, Gérard Galienne.

17. C'est à la fin de mars ou à peu près le temps,
L'année mil neuf cent onze, au début du printemps,
Que nous venons d'apprendre le récit malheureux,
D'un père, son fils, son gendre, et trois autres avec eux

Le matin, on s'empresse, on s'y lève à bonne heure,
Courant à grand' vitesse, au devant du malheur,
On s'en va sur les glaces, marchant, marchant toujours,
On nage dans l'ouverture, jusqu'au milieu du jour.

Environs vers une heure, on trouve les loups-marins,
On les charge à mesure, pour rebrousser chemin,
Mais le vent du contraire qui siffle avec fureur,
Entraîne loin de la terre nos malheureux chasseurs.

C'est vers quatre heures du soir, en vue ils arriviont,
Alors nous autres à terre, s'en retournent aux maisons,
Mais à moitié traverse, un malheureux bouscueuil,
Les soulève, les renverse, et voilà leur cercueil.

Quelle nouvelle attristante il fallait rapporter,
Quelle nouvelle déchirante pour toute la parenté,
Les femmes s'évanouissent et tombent de douleur,
L'esprit, le coeur se brisent, à quoi sert le bonheur?

A quoi sert dans ce monde la gloire et les honneurs?
C'est Dieu, le père du monde, qui dirige nos malheurs.
Allons donc nous soumettre, sa sainte volonté
Et sa main toute-puissante sauront nous protéger.

Avant que de finir, je vais vous les nommer,
L'âme de tous les six que Dieu vient d'appeler:
Il y a Daniel Lebel, son fils, son gendre, son neveu,
Il y a Philias Boudreau, Cyrice Gallant aussi . . .

18. This is the Northeast Atlantic herd, which whelps in the White Sea and migrates in the summer to the boreal waters off Franz-Joseph Land and the east coast of Greenland. Enjoying a certain revival as a result of the War, it is now hunted by the Russians and the Norwegians – in an eminently rational and humane way, it must be added, particularly on the part of the Soviets. Since the whelping ice moves from the White Sea to the Barents Sea, the pups are captured alive while they are moulting, before reaching international waters. They are transported back to land and kept in parks until the moulting process is complete. Then they are slaughtered. In this way, the pinnipeds are exploited not only for their furs, but also for their oil and meat.

19. On file at the Laboratoire de Recherches Arctiques, in Sainte-Anne-de-Bellevue, a suburb of Montreal.